INFORMATION AND COMPUTER SECURITY FOR ACTIVITIES INVOLVING RADIOACTIVE MATERIAL AND FOR ASSOCIATED FACILITIES

The following States are Members of the International Atomic Energy Agency:

AFGHANISTAN
ALBANIA
ALGERIA
ANGOLA
ANTIGUA AND BARBUDA
ARGENTINA
ARMENIA
AUSTRALIA
AUSTRIA
AZERBAIJAN
BAHAMAS
BAHRAIN
BANGLADESH
BARBADOS
BELARUS
BELGIUM
BELIZE
BENIN
BOLIVIA, PLURINATIONAL STATE OF
BOSNIA AND HERZEGOVINA
BOTSWANA
BRAZIL
BRUNEI DARUSSALAM
BULGARIA
BURKINA FASO
BURUNDI
CABO VERDE
CAMBODIA
CAMEROON
CANADA
CENTRAL AFRICAN REPUBLIC
CHAD
CHILE
CHINA
COLOMBIA
COMOROS
CONGO
COOK ISLANDS
COSTA RICA
CÔTE D'IVOIRE
CROATIA
CUBA
CYPRUS
CZECH REPUBLIC
DEMOCRATIC REPUBLIC OF THE CONGO
DENMARK
DJIBOUTI
DOMINICA
DOMINICAN REPUBLIC
ECUADOR
EGYPT
EL SALVADOR
ERITREA
ESTONIA
ESWATINI
ETHIOPIA
FIJI
FINLAND
FRANCE
GABON
GAMBIA, THE
GEORGIA
GERMANY
GHANA
GREECE
GRENADA
GUATEMALA
GUINEA
GUYANA
HAITI
HOLY SEE
HONDURAS
HUNGARY
ICELAND
INDIA
INDONESIA
IRAN, ISLAMIC REPUBLIC OF
IRAQ
IRELAND
ISRAEL
ITALY
JAMAICA
JAPAN
JORDAN
KAZAKHSTAN
KENYA
KOREA, REPUBLIC OF
KUWAIT
KYRGYZSTAN
LAO PEOPLE'S DEMOCRATIC REPUBLIC
LATVIA
LEBANON
LESOTHO
LIBERIA
LIBYA
LIECHTENSTEIN
LITHUANIA
LUXEMBOURG
MADAGASCAR
MALAWI
MALAYSIA
MALI
MALTA
MARSHALL ISLANDS
MAURITANIA
MAURITIUS
MEXICO
MONACO
MONGOLIA
MONTENEGRO
MOROCCO
MOZAMBIQUE
MYANMAR
NAMIBIA
NEPAL
NETHERLANDS, KINGDOM OF THE
NEW ZEALAND
NICARAGUA
NIGER
NIGERIA
NORTH MACEDONIA
NORWAY
OMAN
PAKISTAN
PALAU
PANAMA
PAPUA NEW GUINEA
PARAGUAY
PERU
PHILIPPINES
POLAND
PORTUGAL
QATAR
REPUBLIC OF MOLDOVA
ROMANIA
RUSSIAN FEDERATION
RWANDA
SAINT KITTS AND NEVIS
SAINT LUCIA
SAINT VINCENT AND THE GRENADINES
SAMOA
SAN MARINO
SAUDI ARABIA
SENEGAL
SERBIA
SEYCHELLES
SIERRA LEONE
SINGAPORE
SLOVAKIA
SLOVENIA
SOMALIA
SOUTH AFRICA
SPAIN
SRI LANKA
SUDAN
SWEDEN
SWITZERLAND
SYRIAN ARAB REPUBLIC
TAJIKISTAN
THAILAND
TOGO
TONGA
TRINIDAD AND TOBAGO
TUNISIA
TÜRKİYE
TURKMENISTAN
UGANDA
UKRAINE
UNITED ARAB EMIRATES
UNITED KINGDOM OF GREAT BRITAIN AND NORTHERN IRELAND
UNITED REPUBLIC OF TANZANIA
UNITED STATES OF AMERICA
URUGUAY
UZBEKISTAN
VANUATU
VENEZUELA, BOLIVARIAN REPUBLIC OF
VIET NAM
YEMEN
ZAMBIA
ZIMBABWE

The Agency's Statute was approved on 23 October 1956 by the Conference on the Statute of the IAEA held at United Nations Headquarters, New York; it entered into force on 29 July 1957. The Headquarters of the Agency are situated in Vienna. Its principal objective is "to accelerate and enlarge the contribution of atomic energy to peace, health and prosperity throughout the world".

INFORMATION AND COMPUTER SECURITY FOR ACTIVITIES INVOLVING RADIOACTIVE MATERIAL AND FOR ASSOCIATED FACILITIES

INTERNATIONAL ATOMIC ENERGY AGENCY
VIENNA, 2025

For further information on this publication, please contact:

Information Management Section
International Atomic Energy Agency
Vienna International Centre
PO Box 100
1400 Vienna, Austria
Email: Official.Mail@iaea.org

Printed by the IAEA in Austria
April 2025
https://doi.org/10.61092/iaea.f5kv-z2oz

IAEA Library Cataloguing in Publication Data

Names: International Atomic Energy Agency.
Title: Information and computer security for activities involving radioactive material and for associated facilities / International Atomic Energy Agency.
Description: Vienna : International Atomic Energy Agency, 2025. | Includes bibliographical references.
Identifiers: IAEAL 25-01754 | ISBN 978-92-0-107925-1 (paperback : alk. paper) | ISBN 978-92-0-108025-7 (pdf)
Subjects: LCSH: Nuclear industry — Security measures. | Nuclear facilities — Security measures. | Computer security. | Cyberterrorism.
Classification: 621.039:004.056 | IAEA-TDL-013

FOREWORD

The aim of nuclear security is to prevent, detect and respond to criminal or other intentional unauthorized acts that involve nuclear or other radioactive material and associated facilities. Computers, computing systems and digital components have been playing an ever expanding role in the protection of sensitive information, nuclear material, nuclear safety and security, and material accountancy and control. The advancement of digital technologies has enhanced efficiency and functionality across all nuclear sectors, including those focusing on radioactive material for beneficial purposes. However, it has also introduced new avenues for potential security breaches, necessitating a robust protection strategy for such material throughout its life cycle.

Given the potential for the misuse of radioactive sources in radiological dispersal or exposure devices, protection needs to be ensured for sources that are commonly used in medical, industrial and research settings. With the evolving threat landscape, it is important for interested parties across governments, regulatory bodies, industry and international organizations to coordinate their efforts within a comprehensive nuclear security regime. This coordination entails not only securing physical access to radioactive material but also protecting the digital infrastructure that supports its management and operation.

The integration of digital technologies, including computer based systems and networks, into facilities handling radioactive material involves the implementation of information and computer security measures to prevent unauthorized access, data breaches and attempts at sabotage. This publication addresses these crucial concerns by providing information on practical strategies for implementing information and computer security measures in the context of protecting radioactive material. It underscores the significance of confidentiality, integrity and availability when protecting sensitive digital assets and outlines key nuclear security concepts, including vulnerability, threat, compromise and risk.

The publication also emphasizes the importance of adopting a graded approach tailored to the specific needs of facilities and organizations handling radioactive material. By providing information on best practices and risk mitigation strategies, it aims at equipping interested parties with the necessary tools to fortify their digital defences and ensure the safety and security of radioactive material and associated facilities.

The IAEA is grateful to experts from Australia, Austria, Canada, Germany, the Kingdom of the Netherlands, the Russian Federation and the United States of America for their contributions to this publication, in particular G. Herdes and G. White (United States of America). The IAEA officer responsible for this publication was T. Nelson of the Division of Nuclear Security.

CONTENTS

1. INTRODUCTION

Nuclear security seeks to prevent, detect and respond to criminal or other intentional unauthorized acts involving nuclear material, other radioactive material, associated facilities and associated activities [1]. Radioactive material that is employed for beneficial purposes — used throughout the world in applications such as agriculture, industry, construction, medicine, mining, research and transport — is included in this definition of nuclear security. Nuclear security protects radioactive material throughout its lifetime, from manufacture and use to storage and disposal. Computer security incidents involving radioactive sources and reports of illicit trafficking in radioactive material have raised awareness about the risks associated with radioactive sources that are out of regulatory control, both in terms of safety and security.

States will typically focus their resources for radioactive material on the security of high activity radioactive sources, such as ^{137}Cs, ^{60}Co, ^{141}Am and ^{192}Ir, which are used by associated facilities (e.g. hospitals, universities, sterilization facilities) and by industry. A broad range of organizations are involved in the protection of radioactive material, including governments, competent authorities, regulators, first responders, industry and international organizations, all in a coordinated manner within the State's nuclear security regime. The main collective goal of these organizations is to prevent unauthorized access to radioactive material, which could be used in a radiological dispersal or exposure device. A secondary goal is to prevent the sabotage of equipment that could lead to unexpected radiation exposure of members of the public or to broad contamination.

1.1. BACKGROUND

Although the adoption of digital and network technologies in devices and equipment at facilities that use radioactive material brings technical benefits and efficiencies, it also introduces new classes of threats, attacks and associated risks that could result in the compromise of nuclear security. Computer based systems[1] are devices or equipment that use or are supported by digital technologies (e.g. computers, computing systems, digital components and communication networks). Computer based systems play an important role in physical protection, safety, operations, the management of sensitive information and the maintenance of records of material inventories at facilities, as well as in associated activities involving radioactive material. Information and computer security measures to prevent, protect against, detect[2] and mitigate the consequences of compromise are thus necessary for such computer based systems. Computer systems that perform significant safety or security functions, or that process sensitive information, are also called sensitive digital assets[3], and those that are involved in protecting radioactive material include alarms, access controls, incident response measures and controls for medical devices.

The term 'computer security'[4] is used in this publication to describe "a particular aspect of information security that is concerned with the protection of computer based systems against compromise" [1]. The protection of all interconnected computer based systems and networks is included in this definition of computer security. This publication uses terminology consistent with guidance provided in the IAEA Nuclear Security Series and with the IAEA Nuclear Safety and Security Glossary [1]. Operational technology (OT), industrial control

---

[1] The IAEA Nuclear Safety and Security Glossary [1] defines 'computer based systems' as "Technologies that create, provide access to, process, compute, communicate or store digital information, or perform, provide or control services involving such information", noting that "They may include but are not limited to: desktop, laptop, tablet and other personal computers; mainframe computers; servers; virtual computers; software applications; databases; removable media; digital instrumentation and control devices; programmable logic controllers; printers; network devices; and embedded components and devices."

[2] Paragraph 3.35 of IAEA Nuclear Security Series No. 14, Nuclear Security Recommendations on Radioactive Material and Associated Facilities [2] states: "The *regulatory body* should establish requirements for *operators*, *shippers* and/or carriers to have appropriate and effective security measures to detect *nuclear security events* and to report any such event promptly with the aim of providing a timely response."

[3] Reference [1] defines 'sensitive digital assets' as "sensitive information assets that are (or are parts of) computer based systems."

[4] Other terms synonymous with computer security include 'information technology security', 'information and communications technology security', 'operational technology security', 'industrial control systems security' and 'cyber security'. 'Computer security' is the preferred term in IAEA publications.

systems, and instrumentation and control systems are defined as hardware and software that detect or cause a change through direct monitoring and control of industrial equipment, assets, processes. Examples of OT are physical protection system components (e.g. alarms, access controls, video surveillance systems). Information technology (IT) systems and information and communications technology refers to business systems and networks.

Individuals or groups who are planning to commit a criminal or other intentional unauthorized act involving radioactive material or an associated facility can benefit from having access to sensitive information or to sensitive information assets in relation to the material, the facility or the security measures in place.

IAEA Nuclear Security Series publications that provide guidance on appropriate measures for the identification, classification and securing of sensitive information to achieve effective information security and computer security within the State's nuclear security regime include the following:

— IAEA Nuclear Security Series No. 20, Objective and Essential Elements of a State's Nuclear Security Regime [3];
— IAEA Nuclear Security Series No. 23-G, Security of Nuclear Information [4];
— IAEA Nuclear Security Series No. 42-G, Computer Security for Nuclear Security [5];
— IAEA Nuclear Security Series No. 10-G (Rev. 1), National Nuclear Security Threat Assessment, Design Basis Threats and Representative Threat Statements [6];
— IAEA Nuclear Security Series No. 8-G (Rev. 1), Preventive and Protective Measures against Insider Threats [7];
— IAEA Nuclear Security Series No. 7, Nuclear Security Culture [8];
— IAEA Nuclear Security Series No. 33-T, Computer Security of Instrumentation and Control Systems at Nuclear Facilities [9];
— IAEA Nuclear Security Series No. 17-T (Rev. 1) Computer Security Techniques for Nuclear Facilities [10];
— IAEA Non-serial Publication, Nuclear Security Management for Research Reactors and Related Facilities [11];
— IAEA Non-serial Publication, Computer Security Incident Response Planning at Nuclear Facilities [12];
— IAEA Non-serial Publication, Conducting Computer Security Assessments at Nuclear Facilities [13].

Reference [2] provides nuclear security recommendations on radioactive material and associated facilities, but it does not offer specific guidance for the application of information and computer security to radioactive material and associated facilities.

The computer security guidance provided in this publication is consistent with the IAEA Code of Conduct on the Safety and Security of Radioactive Sources [14], as well as IAEA Nuclear Security Series publications and good practices in industry.

1.2. OBJECTIVE

This publication provides support for the implementation of computer security in Member States, at facilities and in organizations that use, handle or store radioactive material. The objective of this publication is to assist individuals and organizations responsible for the development, implementation and maintenance of information and computer security measures in the context of an effective and comprehensive programme for the protection of radioactive material in manufacture, use, storage and disposal. This publication addresses the radioactive sources outlined in IAEA Nuclear Security Series No. 11-G, Security of Radioactive Material in Use and Storage and of Associated Facilities [15], and radioactive material in need of protection throughout its life cycle.

This publication will describe the information and computer security objectives of confidentiality, integrity and availability, as well as key related nuclear security concepts, such as vulnerability, threat, compromise and risk.

The primarily focus is on how to apply information and computer security principles, including those concerning sensitive information, sensitive information assets [4] and sensitive digital assets [5], for the safety and security of radioactive material and associated facilities.

This publication will also identify guidance that focuses on a graded approach, demonstrating how it can be applied to the protection of sensitive digital assets performing or supporting activities involving radioactive material (see Refs [9–13]). The intention is to provide comprehensive and targeted guidance for personnel who are implementing information and computer security for the protection of radioactive material and associated facilities.

This publication is primarily intended for competent authorities, including regulatory bodies, as well as management and personnel who are working with radioactive material in operations, security, information technology, maintenance and engineering. This publication can also be used by IT and OT vendors, contractors and suppliers, designers and other organizations concerned with the safety and security of facilities using or storing radioactive material.

1.3. SCOPE

This publication covers information and computer security considerations and measures to protect facilities using, handling and storing radioactive material from criminal or other intentional unauthorized acts aimed at the unauthorized removal of material or at sabotage.

It provides good practices for computer systems that support the safety and security of radioactive material. These good practices concern the following:

- Computer security applied to sensitive digital assets within physical protection systems (e.g. access control, video surveillance and alarm monitoring);
- Safety systems (e.g. radiation protection);
- Computer based systems that control the operation of the radioactive source, such as choosing the location, dose and timing of treatment to a patient using a teletherapy device;
- Auxiliary systems that provide or support the security of radioactive material (e.g. electrical power and distribution; IT communication networks; heating, ventilation and air conditioning; fire detection and protection).

Other IT systems can introduce risks to sensitive digital assets (e.g. IT help desk, work control, patient administration management system, personnel accounting). However, the security of these IT systems is not directly covered in this publication, although the publication does indicate how these risks can be considered when designing and implementing computer security measures for sensitive digital assets.

This publication does not provide guidance on safety considerations for sensitive digital assets supporting associated activities using radioactive material. Such guidance can be found in IAEA Safety Standards Series Nos GSR Part 3 and RS-G-1.9, Radiation Protection and Safety of Radiation Sources: International Basic Safety Standards [16]; and Categorization of Radioactive Sources [17].

1.4. STRUCTURE

This publication is divided into seven sections and three annexes. Section 2 presents existing computer security challenges for the security of radioactive material. Section 3 provides general considerations for information and computer security. Section 4 details risk management considerations for information and computer security. Section 5 provides technical details on the protection of digital assets for radioactive material and associated facilities. Section 6 outlines the different elements of an information security management system for organizations and operators. Section 7 examines the sustainability of computer security programmes. The annexes, provide case studies on computer security to address specific concerns that could result from the failure to address information and computer security.

2. COMPUTER SECURITY CHALLENGES FOR THE SECURITY OF RADIOACTIVE MATERIAL

This section provides details on the current state of practices for the physical protection of radioactive material on the basis of the guidance provided in Refs [2, 3, 14, 15].

Reference [14] provides a foundation for the security of radioactive sources, which can be applied to radioactive material, and associated facilities and activities. The IAEA has also published nuclear security recommendations [2] for the development of State regulations for radioactive sources.

The security of radioactive material is the responsibility of the State [3]. The State designates a competent authority as having the main responsibility to regulate activities involving radioactive material within the State. The protection of sensitive information and sensitive information assets is a subject that is addressed in Ref. [3]. Guidance can also be found in Refs [2, 15], which focus on the protection of sensitive information (i.e. information security). The continued adoption of sensitive information within digital assets that use or control radioactive material makes these devices attractive to adversaries, and provides adversaries opportunities to compromise devices so as to achieve their objectives, which could result in radiological consequences.

The risk of criminal or other intentional unauthorized acts in relation to activities associated with radioactive material are those of sabotage and the unauthorized removal of material. The success of acts that result in the unauthorized removal of radioactive material are strongly correlated with the defeat of physical protection systems by outsider or insider adversaries. The defeat of such systems allows undetected unauthorized access to radioactive material. Many functions of physical protection systems are provided through computer based systems, which makes them vulnerable to compromise from cyber-attacks. Such compromise could degrade, alter or disrupt the functions necessary to protect against, detect and respond to unauthorized access to radioactive material.

The State's competent authorities can evaluate and analyse risks and specify associated licence conditions or actions for the organization to address the potential for cyber-attacks. Risks that could lead to other computer security incidents are to be treated in a similar way, depending on the location, the attractiveness of the material and the security concerns of the State.

2.1. PRIORITIZATION OF ORGANIZATIONAL OBJECTIVES

Historically, many facilities prioritized operational objectives on the basis of considerations for safety, giving less priority to considerations relating to security. Similarly, as the application of computer based systems in operations and physical protection systems has grown, considerations relating to computer security have often remained inadequate.

The focus on operational objectives and organizational constraints can lead to choices that are not conducive to computer security, including the following:

- Accessibility of the public to areas and networks, without isolating those networks containing sensitive digital assets that protect radioactive material;
- Reduction of overhead costs through limitations in the allocation of funding or personnel to computer security, including for training and support contracts (e.g. computer security clauses), for technical solutions (e.g. equipment) and for maintenance;
- Reluctance to impose rules on the use of removable media and portable devices (e.g. flash media, smartphones, laptops, tablets);
- Reluctance to update software, or the firmware of security equipment, including anti-malware;
- Reluctance to maintain an inventory of sensitive digital assets with current configurations;
- Liberty of personnel to create ad hoc network designs (e.g. plug and play, no security management oversight, common use of remote access).

These choices not only introduce potential security vulnerabilities that could be exploited by an adversary intent on undertaking unauthorized removal of material or sabotage, but the overall automation of physical protection systems through the application of digital technology also increases the potential for compromise by remote adversaries' intent on committing criminal or other intentional unauthorized acts.

2.2. A FOCUS ON PHYSICAL PROTECTION SYSTEMS

The guidance in Ref. [15] provides a graded approach to security and the protection of information — through use of physical protection systems — but does not include computer security for the computer based systems and sensitive digital assets that maintain and operate these physical protection systems and the crucial processes that protect all radioactive material.

The State's competent authority requires that organizations develop and implement a security plan[5], which addresses physical protection, without necessarily addressing information and computer security.

3. GENERAL CONSIDERATIONS FOR INFORMATION AND COMPUTER SECURITY

Computer security is concerned with the protection of digital assets within a nuclear security regime against compromise (e.g. cyber-attacks, manipulation, falsification). Information and computer security is a cross-cutting principle within the nuclear security regime, which is important in identifying and mitigating computer security issues that might affect crucial functions and that might prevent processes and computer based systems from operating in a safe and secure manner. Integrating information and computer security into the security plan will thus ensure that adequate protection is provided for the sustainability of nuclear security.

Paragraph 3.3(g) of Ref. [3] states that the legislative and regulatory framework: "Provide[s] for the establishment of regulations and requirements for protecting the confidentiality of *sensitive information* and for protecting *sensitive information assets.*"

Paragraph 3.33 of Ref. [2] states that "The *regulatory body* should ensure that the *operator*'s security plan includes measures to effectively detect, delay and respond to a *malicious act* consistent with the *threat.*"

Paragraph 3.13 of Ref. [5] states that "The State should require the identified competent authorities and operators to develop and implement CSPs [computer security programmes]".

Such programmes ensure the protection of the confidentiality, integrity and availability of sensitive information and sensitive information assets.

Paragraph 3.34 of IAEA Nuclear Security Series No. 43-T, Security Management of Radioactive Material in Use and Storage and of Associated Facilities [18] states that "The security plan enables operators to demonstrate to the regulatory body their compliance with security requirements."

The security plan provides a guide for security personnel at the facility concerning the operation, maintenance and continuous improvement of the security system. Specific elements in relation to information and computer security are to be included in this security plan (see Part II). Detailed guidance is provided on the format and content of the operator's security plan in section 5 of Ref. [18].

The risk of criminal or other intentional unauthorized acts directed at sensitive information or sensitive information assets that are associated with activities involving radioactive material need to be addressed in the regulatory programme and within an organization's security plan.

---

[5] For some facilities, it is called a 'site security plan'.

3.1. SENSITIVE INFORMATION, SENSITIVE INFORMATION ASSETS AND SENSITIVE DIGITAL ASSETS

Facilities that use, store and handle radioactive material also create, process and store many types of information, and this information will have different sensitivity and protection needs. Sensitive information within a nuclear security regime is defined as "information, in whatever form, including software, the unauthorized disclosure, modification, alteration, destruction, or denial of use of which could compromise nuclear security." [1] In turn, this could support the unauthorized removal or sabotage of radioactive material and/or the sabotage of associated facilities.

The relationships between a State's nuclear security regime, information, sensitive information, sensitive information assets, sensitive digital assets and computer based systems is shown in Fig. 1.

Sensitive information at associated facilities could include security plans, radioactive material inventory lists, transport and movement schedules and routes, details on security measures and documentation associated with personnel trustworthiness checks.

Given the widespread and increasing adoption of computer based systems, a significant and growing proportion of sensitive information assets are also considered to be sensitive digital assets. For instance, radioactive material inventory lists are increasingly stored as electronic databases on servers.

Components of physical protection systems (e.g. network cameras, access control devices, network video recorders, operator workstations, network communications equipment) and the digital controllers used in medical devices, as well as other process controls that use radioactive material, are examples of sensitive digital assets in associated facilities. These devices could also have graphical user interfaces, digital information storage and processing, and associated networks.

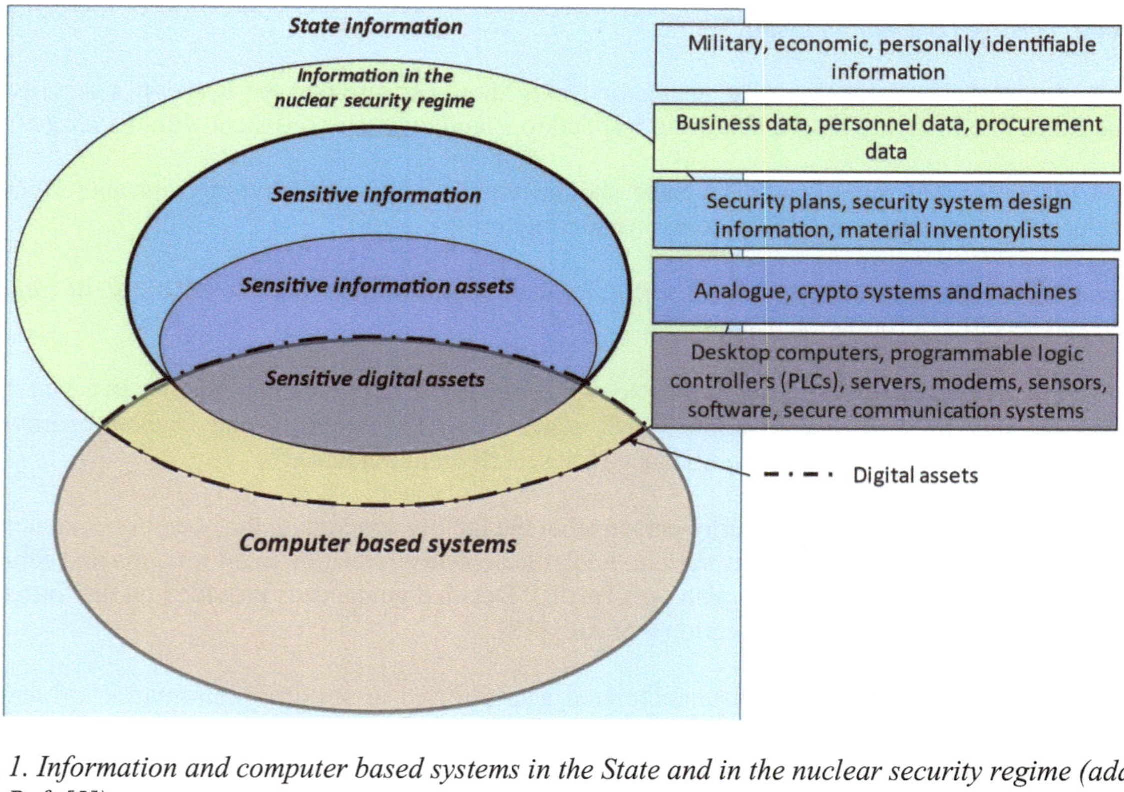

FIG. 1. Information and computer based systems in the State and in the nuclear security regime (adapted from Ref. [5]).

Computer security protection demands are determined on the basis of the potential consequences of the compromise of functions performed by the sensitive digital assets. Examples of functions in a physical protection system include video surveillance and detection of unauthorized access for response capabilities. Such functions would require protection in terms of confidentiality, integrity and the availability of the sensitive digital asset, such as network based camera and/or a database server that contains sensitive information (e.g. recordings, access controls, door controls, sensitive information).

Special consideration needs to be given to cloud based services through 'infrastructure as a service'. Infrastructure as a service could be physical protection systems, patient and care information systems, and material control and accounting systems. These systems transfer the control of some liability to a third party host provider. The organization could specify computer security contractual requirements for compliance by third parties in order to mitigate risks.

3.2. CONFIDENTIALITY, INTEGRITY AND AVAILABILITY

The term 'information security', refers to the system, programme or set of rules in place to ensure preservation of the confidentiality, integrity and availability of information in any form [1].

Physical protection systems, for example, emphasize the integrity and availability of information to ensure a timely response, while patient care systems emphasize confidentiality to protect sensitive medical information from being disclosed. Para. 2.11 of Ref. [4] states:

> "Loss of integrity or availability can negatively affect nuclear security just as loss of confidentiality can. For example, if authorized users do not have timely access to information necessary for their duties (loss of availability), or if that information has been altered in such a manner as to mislead them (loss of integrity)."

Such actions could result in response times being delayed, which could increase the chances of criminal or other intentional unauthorized acts being successful.

Confidentiality is defined as "The property that information is not made available or disclosed to unauthorized individuals, entities or processes" [1]. The disclosure of personal, identifiable information is therefore an example of a loss of confidentiality since this information needs to be protected from unauthorized access.

Integrity is defined as "The property of accuracy and completeness of information" [1]. The loss of integrity of inventory lists of radioactive material is one example of a loss of integrity since it would leave the competent authority with inaccurate information concerning the location, quantity and/or type of radioactive material at facilities, and could impede the timely detection of missing radioactive material.

Availability is defined as "The property of being accessible and usable upon demand by an authorized entity" [1]. A successful cyber-attack on a network camera, for example, would impede communications between the camera and the central alarm station, causing a loss of situational awareness and a potential delay in response.

3.3. EVALUATION OF THE SENSITIVITY OF DIGITAL ASSETS

Digital assets perform or support many functions across a State's nuclear security regime. Assets that are crucial to achieving nuclear security or safety objectives are sensitive digital assets. Some digital assets can provide valuable business and communication functions, but they are not considered sensitive in relation to nuclear security.

Sensitive digital assets need to be evaluated in order to determine the level of protection necessary. This evaluation considers how a compromise of the digital asset affects functions supporting nuclear security objectives, along with the significance of those functions. It also needs to consider the sensitivity of the information being stored, transmitted, processed or accessed by the digital asset.

Targeted cyber-attacks have the potential to compromise functions and sensitive information that are highly significant to nuclear security or safety. For an evaluation to be performed on the basis of functions, the consequences of the compromise of the functions performed by the sensitive digital assets need to be determined. Reference [10] indicates the potential consequences on a function that would arise from the compromise of the system (or digital asset). Specifically, para. 4.22 of Ref. [10] states:

> "These effects are as follows (arranged from worst to best cases):
>
> (a) The performance of the facility function is indeterminate. This means that the function might be altered in any manner without the initial compromise being detected.
> (b) The performance of the facility function changes in unexpected ways (and other actions can be performed), but these anomalies are observable to the operator.
> (c) The performance of the facility function fails.
> (d) The performance of the facility function is as expected, meaning the compromise does not adversely affect the facility function (i.e. the system is fault tolerant)."

From the classical approach to information security, the impacts of disclosure that need to be considered are the following:

— Serious damage, for example unauthorized disclosure of the entire contents of the security plan, detailing all of the measures implemented, as well as vulnerabilities and weaknesses on-site;
— Damage, for example unauthorized disclosure of access control lists for a site's physical protection system;
— An increased difficulty in the provision of security, for example unauthorized disclosure of details in relation to personnel who have responsibilities associated with nuclear security.

The higher the consequence of compromise and/or the sensitivity of the information, the higher the protection associated with the sensitive data asset. Paragraph 6.17 of Ref. [4] introduces a list of information security measures (e.g. physical, administrative, technical controls) that can be considered for implementation, and states:

> "Information security measures include, among other things:
>
> (a) Administrative management to govern, maintain and develop information security (including third party services);
> (b) Personnel security, particularly in the phases of recruiting, and the beginning and end of employment;
> (c) Physical security of areas where sensitive information or sensitive information assets are used, handled or located;
> (d) Security of digital and manual information handling: workstation security, virus and malware protection, deletion and destruction of information, and manual processes;
> (e) Communication network security (telephones, email, the Internet and local area networks): policy, user authentication, equipment identification, segregation, connection and routing controls, and monitoring;
> (f) Equipment security: access control, logging of use, spare part management, backup of critical equipment, backup power arrangements, documentation and maintenance, cabling and media security;
> (g) Software security: access control, logging of user and super user activities, backup management, maintenance contracting, configuration and version management, use of registered, legal software, testing for vulnerabilities and testing for system behaviour under error conditions;
> (h) Security of use of information systems: user rights control, user recognition and verification, connecting to services, systems and equipment, password management, oversight of use, and the two person rule (i.e. two person control) for critical operations;
> (i) Classification and corresponding procedures for handling information;
> (j) Protection of privacy."

3.4. INTERFACES BETWEEN COMPUTER SECURITY AND OTHER SECURITY DOMAINS

It is important to apply a holistic approach to information and computer security (i.e. an approach that is not performed in isolation from other security domains) to ensure that risk reduction measures are maintained and effective. The subsections below describe the key interfaces between computer security and other security domains.

Support for computer security could be ensured through the promotion of effective interfaces between personnel in IT and OT (including personnel working on physical protection systems) since the underlying domains are co-dependent. IT personnel need to understand the OT systems and networks that are integrated within the organization's overall architecture, while OT personnel need to understand the strengths and weaknesses (and the lack of security) within the OT architecture, which could have an impact on dedicated systems (i.e. implementations of computer security controls on OT systems without a knowledge of the device's functions and operating parameters could adversely impact the systems and security).

3.4.1 Interface with physical protection

Physical protection personnel will need to interface with the following individuals or groups of individuals:

- The IT group, to ensure that communication from physical protection systems to monitoring systems are protected.
- Computer security, which could be separate from or part of the IT group, to identify and develop policies and procedures that support the effective operation of computer based systems within the physical protection system and that ensure compliance with regulations.
- The security equipment vendor or installer, and the entity maintaining sensitive data assets (e.g. security components, applications), to ensure that computer security processes, such as hardening, are effectively implemented.
- Personnel responsible for communications technologies used in coordinating the on- and off-site security response, to ensure protection against security threats.
- The training organization providing computer security training. All personnel need to be aware of and trained in basic computer security awareness (e.g. informed not to click on phishing links), and personnel having specific security responsibilities need to be provided role based training.

3.4.2 Interface with safety

Activities associated with the safety–security interface are outlined in Ref. [15] and generally involve digital assets. Information and computer security assessments performed by competent personnel using mature organizational processes are necessary to ensure the safety and security of radioactive material.

3.5. THE INFORMATION AND COMPUTER SECURITY PROGRAMME

The computer security programme defines the organization's role in implementing its strategy, for example in terms of organizational roles, responsibilities, policies and procedures. Security procedures need to be implemented and maintained throughout the lifetime of the facility, including procedures that concern sensitive information and sensitive digital assets. The potential effects resulting from the modification of sensitive information and sensitive digital assets are to be outlined in these security procedures. Procedures also need to identify actions that are to be undertaken in response to the compromise of assets.

The computer security programme specifies how the organization aims to achieve computer security objectives and perform the computer security measures specified by legislation, regulations, standards and guidance from the regulatory body and from the competent authority for computer security.

Reference [5] provides guidance on the overall means for implementing computer security programmes, and Ref. [10] gives examples of how these elements can be applied to a computer security programme for a nuclear power plant. Specific guidance from these publications, adapted for radioactive material, is provided in Sections 5 and 6 of this publication.

Paragraph 3.34 of Ref. [18] states:

> "The security plan enables operators to demonstrate to the regulatory body their compliance with security requirements. A security plan is an important tool for documenting the activities associated with establishing, implementing and maintaining an effective, sustainable and integrated security system."

Some elements of an organization's information and computer security programme are likely to already exist within the organization's security plan, as required by the State's regulatory framework. These elements will largely focus on the protection of the confidentiality of sensitive information, but protection of integrity and availability is an important and necessary consideration. Reference [4] provides detailed guidance on developing a framework for information security.

Specific considerations for the inclusion or modification of the security plan in relation to information and computer security are provided in Section 6 of this publication.

3.6. PRACTICAL IMPLEMENTATION OF THE COMPUTER SECURITY PROGRAMME

Computer based systems introduce inherent vulnerabilities, increasing the potential for cyber-attacks, which need to be addressed in order to ensure ongoing nuclear security and safety. Implementing all of the elements of a computer security programme will depend on available financial and human resources. Consequently, the implementation of a computer security programme could be undertaken progressively, beginning with an initial stage that could provide reasonable assurance of nuclear security and safety. This initial implementation stage would form the basis for the continuous improvement of computer security through increasingly mature stages of implementation aiming to deliver the full extent of the computer security programme. Table 1 presents an example of how the elements of a computer security programme could be aligned with these different stages of implementation (i.e. initial, intermediate, advanced). The data in Table 1 are adapted from Refs [9, 19], as well as from Section 5 of this publication.

3.7. COMPUTER SECURITY LEVELS — A GRADED APPROACH

Paragraph 4.3 of Ref. [2] states that "Security requirements for *radioactive material* should be based on a *graded approach*."

For computer security, a graded approach can be represented by the concept of computer security levels [10]. Computer security levels provide protection to digital assets that support the nuclear security of radioactive material.

The level of computer security for the digital asset is associated with a set of security objectives, the stringency of which are strongly correlated to the severity of the associated potential consequence resulting from the compromise of the system function or the unauthorized disclosure of the sensitive information. The greater the risk or severity of the consequence, the stricter the protection measures. These measures, once imposed, increase the level of assurance in relation to the security of sensitive digital assets.

A computer security level is a designation that indicates the degree of security protection necessary for a function and for the sensitive information, and consequently, for the digital system that performs that function or that accesses, stores, controls or transmits the information. Each computer security level is associated with a set of requirements imposed by the operator to ensure that the appropriate level of protection is provided to digital assets assigned to that level, on the basis of a graded approach. Each computer security level will need different sets of computer security measures to satisfy the computer security objectives for that level. More detailed guidance for nuclear facilities is provided in Refs [5, 10]. For radioactive material, however, specific objectives for computer levels, and for what systems are to be assigned to each level, are to be specified in guidance provided by the competent authority in coordination with the facility.

The primary attribute of a computer security level is that the established set of security objectives leads to graded requirements and to conditions being imposed on facility processes, systems and activities. These requirements and conditions are 'designed' to mandate the application of the greatest degree of resources to those functions — and the systems performing them — that are associated with the most severe consequences, if compromised.

TABLE 1. IMPLEMENTATION GUIDE: COMPUTER SECURITY PROGRAMME MATURITY LEVEL

Element of the computer security programme	Reference	Implementation level		
		Initial	**Intermediate**	**Advanced**
System security design and configuration management: Fundamental architecture and design principles	Implementation of computer security boundaries. (Ref. [9], para. 4.177)	A boundary (i.e. either logical or physical) is defined and implemented in the design stage.	Logical and physical boundaries are known or documented, with protection provided commensurate with the risk.	Logical and physical boundaries are tightly coupled, and have common and supportive boundary protection measures.
Personnel management: Awareness raising and training	Training considerations for personnel performing work. (Ref. [9], para. 4.54)	A computer and information security awareness module is made compulsory for all personnel before they are permitted to perform work involving sensitive data assets.	After completion of the initial stage, a training programme is established that is better able to target training to specific roles, with periodic refresher training.	After the initial and intermediate stages, specialists then receive dedicated external training on industry best practices in relation to computer security procedures.
Operational security procedures: System backup	Backup and restoration for contingency planning. (Ref. [9], para. 4.45) (Ref. [19], no. 15)	Backups are taken and restoration is performed on a periodic basis.	After completion of the initial stage, backups of software, essential data and configuration files are maintained in a secure secondary location that can reduce ‘time to recovery’ or re-creation. Security measures (e.g. cryptography, physical protection) are in place to protect backups from theft, tampering, deletion or destruction.	After the initial and intermediate stages, the functionality of the digital asset environment is known to be restorable in a predictable time frame.

TABLE 1. IMPLEMENTATION GUIDE: COMPUTER SECURITY PROGRAMME MATURITY LEVEL (cont.)

Element of the computer security programme	Reference	Implementation level		
		Initial	**Intermediate**	**Advanced**
Operational security procedures: Access control	Measures to prevent unauthorized access. (Ref. [9], para. 4.200)	Access to sensitive digital assets is controlled using individual authentication.	After the initial stage, access to sensitive digital assets is controlled using multifactor authentication, where practical. Both physical and technical control measures are implemented to prevent unauthorized access.	After the initial and intermediate stages, tamper indicating devices are implemented to detect unauthorized physical access to sensitive digital assets. Access to computer based systems is granted with the least privileges (i.e. the minimal set of access and capabilities necessary to perform authorized activities).
Asset management: List of all digital assets	An inventory is maintained of digital assets, including software, subsystems and components. The list is updated on a regular basis. (Ref. [9], para. 3.23)	Digital assets associated with the physical protection system are inventoried and placed under computer security management. Information associated with the physical protection system or the radioactive material, and relevant to computer security, is analysed to determine the level of sensitivity of the information.	After the initial stage, digital assets associated with radioactive material (e.g. operational technology) or a supporting physical protection system, as well as radioactive material processes and sensitive information, are inventoried and are placed under computer security management.	

3.8. COMPUTER SECURITY ZONES FOR DEFENCE IN DEPTH

Computer security zones, and associated attributes, are important building blocks in a computer security strategy and a crucial aspect of a defensive computer security architecture.

Paragraph 2.9 of Ref. [10] states:

> "A computer security zone is a logical and/or physical grouping of digital assets that are assigned to the same computer security level and that share common computer security requirements owing to inherent properties of the systems or their connections to other systems (and, if necessary, additional criteria). The use of computer security zones is intended to simplify the administration, communication and application of computer security measures."

Examples of computer security zones for radioactive material might include a patient information system, a physical intrusion detection system or an access control system, each of which would be placed in its own zone.

Because of the interconnectivity of digital assets and associated networks, including the information flow, computer security measures need to be implemented to protect sensitive digital assets from attacks that could originate from other digital assets to which they are connected. Defence in depth for computer security entails the establishment of zones and their arrangement, in accordance with similar functions, within a defensive computer security architecture, with a graded application of security measures [10]. Such an architecture would have objectives established for the different levels of security, with the strength of computer security controls or measures increasing in accordance with the different levels that are implemented in each computer security zone. This combination of successive layers of computer security creates a defence in depth architecture that would have to be overcome or bypassed by an adversary to compromise any function or system.

Additional criteria for defining computer security zones could include the following:

- Underlining organizational responsibilities (e.g. the computer security zones could differ for systems that have different departments responsible for them);
- Maintaining separation (e.g. different computer security zones are defined for redundant systems that are at the same computer security level and are performing the same function);
- Using zones defined for other purposes (e.g. for simplicity, a computer security zone could be defined as being the same as a zone already established for administrative or communication purposes).

The design of the physical protection system needs to incorporate information concerning zones in which to deploy independent measures so that the failure of one measure does not mean the compromise of digital assets within the zone. Such independent measures could include physical, administrative and technical measures. A physical measure, for example, might comprise multiple independent and diverse physical barriers (e.g. hardened buildings, hardened doors, cages, tamper indicating devices, tie-downs). Administrative measures could include policies and procedures for personnel access, incidental use, portable media and mobile device use, or procurement language. Technical measures might consist of network or host intrusion detection systems, firewalls, virtual local area networks or antiviruses.

3.9. REGULATORY CONSIDERATIONS

The competent authority could consider introducing licence conditions that include information and computer security. Such licence conditions need to be supported by an effective inspection programme, with inspectors competent in computer security and with some type of corrective or enforcement actions — up to and including licence revocation — if deficiencies are discovered.

The competent authority could implicate other organizations with capabilities and competence in computer security. These organizations could include the national computer security centre, technical support organizations (e.g. nuclear safety, computer security emergency response team) or other competent authorities and intelligence services.

The competent authority needs to define computer security requirements and regulations for the appropriate levels of protection, using a risk informed approach that undertakes the following:

- Ensures that requirements reflect the State's information and computer security strategy, policy and the requirements of the appropriate relevant entity (e.g. including the necessary capabilities and competences);
- Prescribes specific computer security measures for the competent authorities or operators to implement in accordance with the risk assessment (e.g. a prescriptive approach);
- Verifies continued compliance with computer security requirements through the conduct of regular assurance activities (e.g. assessments, evaluations, audits), and when necessary, ensuring that corrective actions are taken.

Paragraph 4.18 of Ref. [5] states:

> "The criteria for the selection of a prescriptive approach or a performance based approach (or an appropriate combination of the two) will depend on the State's legislative framework and organizational structure and several other factors such as the following:
>
> (a) The competence of the operator to interpret performance requirements and to design, implement and evaluate an effective nuclear security system;
> (b) The number and variety of different facilities and operators that will be governed by the regulation, and the extent to which prescriptive requirements might limit the flexibility of the operator to develop appropriate measures;
> (c) The severity of the potential consequences of the malicious acts that are to be prevented or protected against".

The competent authority needs to ensure that computer security regulations are adaptable to changes in computer based systems, for example in the case of changes in threats and in the nature of cyber-attacks, and in the case of available updates in computer security measures.

Reference [5] provides more detailed information on the prescriptive, performance or combined approaches.

4. RISK MANAGEMENT IN INFORMATION AND COMPUTER SECURITY

Information and computer security contributes to safety and security in all nuclear domains (e.g. at nuclear facilities, in the case of radioactive material and associated facilities and activities; and in the case of material outside of regulatory control), through a risk informed approach, as stated in Essential Element 9 of Ref. [3]:

> "A *nuclear security regime* uses risk informed approaches, including in the allocation of resources for *nuclear security systems* and *nuclear security measures* and in the conduct of nuclear security related activities that are based on a *graded approach* and *defence in depth*, which take into account the following:
>
> (a) The State's current assessment of the *nuclear security threats*, both internal and external;
> (b) The relative attractiveness and vulnerability of identified *targets* to *nuclear security threats*;
> (c) Characteristics of the *nuclear material, other radioactive material, associated facilities* and *associated activities*;
> (d) Potential harmful consequences from criminal or intentional unauthorized acts involving or directed at *nuclear material*, *other radioactive material*, *associated facilities*, *associated*

activities, *sensitive information or sensitive information assets*, and other acts determined by the State to have an adverse impact on nuclear security."

4.1. RISK ASSESSMENT IN INFORMATION AND COMPUTER SECURITY

Paragraph 3.11 of Ref. [10] states:

> "Risk, in the computer security context, is the risk associated with an adversary exploiting the vulnerabilities of a digital asset or group of digital assets to commit or facilitate a malicious act. The risk is expressed as a combination of the likelihood of a successful attack and the severity of its consequence if it occurs."

Computer security risks in relation to the security of radioactive material will be associated with one of the following two types of nuclear security event:

(1) Criminal or other intentional unauthorized acts leading to the unauthorized removal of radioactive material or to radioactive contamination (e.g. compromise of physical protection system equipment or of a safety system);
(2) Criminal or other intentional unauthorized acts resulting in sabotage of radioactive material, facilities and associated activities (e.g. compromise of devices resulting in unintended radiation exposure to personnel or the public).

Paragraph 3.15 of Ref. [4] states that "The recommended way of assessing the value of a particular information asset is to use a risk informed approach, considering the...consequences that are likely to occur in the event of its compromise."

To support a risk assessment, a list of relevant nuclear security scenarios needs to be developed with the associated types of attack (e.g. cyber, physical, blended) and their potential consequences. For example, the effects of ransomware attacks, the disruption of facility operations, or the theft of intellectual property could serve as a basis for developing scenarios that could lead to unauthorized removal or sabotage. The best practices for computer security that are outlined in Ref. [19] could be assessed when developing these scenarios, including consideration of the following information:

— Cyber-attacks can be undertaken to manipulate or sabotage equipment and processes that use radioactive sources.
— An adversary could exploit security equipment to gain access to a site's network(s) in order to perform a cyber-attack, for example by installing ransomware or stealing proprietary or other sensitive information.
— Social engineering could be used to exploit unwitting insiders in order to gain access to physical protection systems, networks and related subsystems without the need to hack or conduct a cyber-attack using cyber tools.

4.1.1 Vulnerabilities

A computer security vulnerability is a defect or a weakness in a digital asset, or a computer security measure that can be exploited by an adversary. Computer security measures might have been adopted in an ad hoc manner in the past, or simply not considered in the design and implementation of physical protection systems, resulting in systems and networks with vulnerabilities. All physical protection systems and digital assets are likely to have vulnerabilities, including vulnerabilities that exist at the facility level and those involving individual digital assets alone. Examples of facility level vulnerabilities include the following:

— Flat networks (i.e. no segmentation);
— Lack of defined security responsibilities;
— Lack of role based training for personnel;
— Ineffective or absent security measures;

— Inappropriate or inadequate administrative controls;
— Inadequate communication protection;
— Poor security culture.

Examples of vulnerabilities involving a digital asset include the following:

— Improper input validation;
— Use of hard-coded credentials;
— Incorrect permission assignments;
— Improper authentication.

Vulnerabilities that could potentially have an effect on functions and contribute to increased risks can be identified at facilities with computer systems. Information concerning vulnerabilities and their severity can be found in publicly available data sources, such as the common vulnerabilities and exposures (CVE) list[6], and other sources such as 'CVE details', which is a database that is sorted according to vendors, products and vulnerability type[7]. Research on digital assets can be undertaken using open source resources published by professional organizations in the computer security industry.

One commonly used resource that provides a scoring scheme in relation to the severity of vulnerabilities is the Common Vulnerability Scoring System (CVSS)[8]. The score approximates the ease of use and impact of an act to assist facilities with risk management.

A physical protection system that has one or more crucial vulnerabilities results in the entire system being unable to protect against a targeted and sophisticated attack. Vulnerabilities thus increase the likelihood of a compromise to the physical protection system function(s) and increase the risk associated with unauthorized removal of radioactive material and the risk associated with sabotage.

4.1.2 Likelihood

Reference [20] defines 'likelihood' in the context of information and computer security management as "the chance of something happening", which is synonymous with 'probability'. Computer security scenarios need to be assessed — either qualitatively or quantitatively — to determine the likelihood of incidents occurring. Such a likelihood assessment could consider the following factors, taken from Ref. [21]:

— "Experience and applicable statistics for risk source likelihood;
— For deliberate risk sources: the degree of motivation [e.g. the viability (cost/benefit) of the attack] and capabilities (e.g. the level of the skill of possible attackers), which change over time, resources available to possible attackers, and influences on possible attackers such as serious crime, terrorist organizations or foreign intelligence, as well as the perception of attractiveness and vulnerability of information for a possible attacker;
— For accidental risk sources: geographical factors (e.g. proximity to dangerous facilities or activities), the possibility of natural disasters such as extreme weather, volcanic activity, earthquakes, flooding, tsunami and factors that can influence human errors and equipment malfunction;
— Known weaknesses and any compensating controls, both individually and in aggregation;
— Existing controls and how effectively they reduce known weaknesses."

Many States do not consider the likelihood of computer security incidents in relation to radioactive material, but the use of scenarios could greatly assist in the identification and prioritization of risks.

[6] For more information, see: https://cve.mitre.org
[7] For more information, see: www.cvedetails.com
[8] For more information, see www.first.org/cvss/

4.1.3 Severity of consequences

The compromise of sensitive data assets could lead to a loss of the safety and/or security of radioactive material and associated facilities. The severity of the consequences is associated with the categorization of the radioactive material or the classification of the sensitive information. Such categorization ensures that greater levels of protection are implemented for radioactive material associated with higher consequences in the case of a criminal or other intentional unauthorized act being successful. The aggregation of radioactive material could be taken into account in this categorization system. The level of protection of sensitive information is determined on the basis of the consequences in the case of disclosure or loss.

4.1.4 Cyber-attacks and blended attacks

A cyber-attack is defined as "a malicious act with the intention of stealing, altering, preventing access to or destroying a specified target through unauthorized access to (or actions within) a susceptible computer based system" [1].

Cyber-attacks are the means by which an adversary exploits vulnerabilities that are present in the system. The adversary's goal is to compromise the system in order to degrade, disrupt or destroy its ability to perform nuclear security functions. The goal of the adversary could also be to support a physical attack, which might result in unauthorized or undetected access to radioactive sources.

Given the adoption of digital assets in devices that use or control radioactive material, cyber-attacks are a growing threat to the nuclear security of radioactive material. Examples of cyber-attacks could include:

- Unauthorized and unobservable modifications of patients' treatment profiles (e.g. dose rate, target location) on a medical device through compromise of the digital logic.
- Unauthorized modification of digital safety interlocks (e.g. shielding structure, safety alarm setpoints, radiation monitoring infrastructure) resulting in unintended exposure to a high activity source.
- Compromise of digital control systems used in the production of radioactive sources or the processing of radioactive waste, resulting in a radioactive release to the environment beyond licence limits.

Modern physical protection systems employ digital assets that could provide opportunities for adversaries to compromise these systems. The tactics, techniques and procedures of adversaries can include use of cyber capabilities in the initial stages of an attack (e.g. to conduct reconnaissance) and during the attack (e.g. to degrade the performance of the physical protection system or delay and disrupt the response). Attacks that have been associated with the highest risk in relation to the unauthorized removal of radioactive material are blended attacks (i.e. attacks that have both a cyber component and a physical component).

Blended attacks could involve the compromise of the physical protection system through a cyber-attack to disable, disrupt or degrade security functions prior to the initiation of, and during all phases of, the physical element of the attack, which could include delaying or preventing a response by law enforcement or other response agencies. Such blended attacks will increase the probability that the attack will be successful since the physical protection system will be unable to detect, trigger an alarm or prevent access to the material.

4.2. ADVERSARIES AND CYBER CAPABILITIES

Cyber skills can be developed within an adversary group or acquired as a capability from third parties (e.g. criminal organizations or hacker groups). Advanced hacking tools can be freely obtained online, and social engineering can also be used to facilitate a cyber-attack.

Reference [7] defines two types of adversaries: (1) external adversaries, who do not have authorization or access to crucial equipment or radioactive material at the site; and (2) insider adversaries, who have authorization and access to crucial equipment or radioactive material on the site. The collusion of an external adversary with an insider adversary is particularly effective in achieving a criminal or other intentional unauthorized act.

The unwitting insider within an organization is a major concern since a number of personnel have privileged and authorized access to sensitive areas and equipment. An unwitting insider is usually compromised through social engineering techniques, such as a phishing email that allows access to networks, user accounts and devices. The unwitting insider is unaware of the compromise and consequently does not exhibit behavioural changes that could be observed by other personnel or by management.

Large amounts of open source information are potentially available on organizations, personnel and facilities that use radioactive material, and this information could be used by adversary groups when planning a cyber-attack.

The operator needs to be provided with threat information from the competent authority or the State in sufficient detail to indicate how the physical protection system is to be designed to protect against both external and internal threats. The operator needs to indicate to the competent authority who is responsible for receiving threat information and how such information is to be shared with personnel on a need to know basis.

4.3. RISK MANAGEMENT IN INFORMATION AND COMPUTER SECURITY

4.3.1 Prioritization of resources

Targeted attacks are those most likely to result in nuclear security events. An effective approach for the computer security of an organization is to implement general computer security measures to prevent incidents of lower consequence, while focusing on the implementation of specific computer security measures to prevent targeted attacks that could lead to unauthorized removal and/or sabotage of radioactive material from succeeding.

Implementation of computer security measures in this manner will provide effective computer security while at the same time ensuring that compliance costs remain manageable. It prioritizes available resources to implement computer security measures in the most effective manner, using a graded approach.

In many States, the competent authority issues regulatory requirements that are used by the operator as a basis for justifying expenditures on nuclear security. The State strategy therefore needs to ensure that the necessary resources (e.g. financial, human, technical) are made available to the competent authority so as to provide computer security for radioactive material and associated facilities [5].

Organizations also need to be in a position to allocate sufficient financial and human resources in order to comply with the regulatory requirements of the State. For example, organizations need to ensure delivery of the different elements of the security plan, such as investment in qualified personnel or contractors, systems, components, equipment and the tools for risk reduction.

4.3.2 Capability and maturity

To support effective computer security, the State strategy [5] needs to enable the development of qualified computer security professionals to support protection against unauthorized removal and sabotage of radioactive material and associated facilities. Competent personnel have the necessary skills to support compliance with regulations.

Organizations need to develop capabilities and maturity through investment in the scope of the programme, assets and resources. Such capability development can be supported through the adoption of a maturity model[9]. Maturity models focus on the progression and evolution of organizational processes that continually improve and adapt to new security challenges and objectives [22].

---

[9] Examples of maturity models are the US Department of Energy's Cybersecurity Capability Maturity Model (C2M2) at www.energy.gov/ceser/cybersecurity-capability-maturity-model-c2m2 and the NIST Cybersecurity Framework at www.nist.gov/cyberframework

4.3.3 Competence

It is important that personnel be trained and supported to develop the appropriate competencies that are based upon knowledge, skills and abilities. Specialized resources need to be made available on demand for information and computer security, which entails an investment in both internal and external human resources. This investment is to be directed in a purposeful way to ensure the progression of competence among personnel. The training programme for information and computer security needs to be developed in accordance with the maturity model that is employed by the organization.

Mature organizations with competent personnel are those that are most well suited to ensure information and computer security for the protection of radioactive material and associated facilities. In the case of organizations with limited resources in relation to computer security knowledge or capabilities, contractual arrangements with competent computer security professionals could be necessary.

4.3.4 Procurement of services and equipment

Specific contractual conditions for information and computer security are to be included in third party contracts for the procurement of services and equipment. Key considerations for contractual conditions include the following:

- Mandatory trustworthiness checks for third party personnel and organizations having access to, or knowledge of, sensitive digital assets;
- Provisions to allow for the assessment of vendor computer security programmes;
- Acceptable use policies for assets and technology;
- Support for computer security (e.g. reporting, provision of security updates or patches, assistance during incident response);
- Conditions for information transfer and use (see Clause 5.14 Ref. [23]);
- Computer security approaches used in the design of equipment and software, such as a secure development policy, secure system engineering principles, or a secure development environment (see Clause 8.25 of Ref. [23]);
- An information security policy for suppliers (see Clause 5.19 of Ref. [23]).

Reference [24] provides a set of suggested procurement language that is applicable to physical protection systems for radioactive material. Such procurement language could be adapted to third party contracts with physical protection system suppliers. The key sections of this publication are the following:

- Software and services;
- Access control;
- Account management;
- Session management;
- Authentication/password policy and management;
- Logging and auditing;
- Communication restrictions;
- Malware detection and protection;
- State of health signals;
- Reliability and adherence to standards;
- Documentation and tracking of vulnerabilities;
- Problem reporting;
- Patch management and updates;
- Host intrusion detection;
- Network intrusion detection;
- Cryptographic system documentation;
- General wireless technology provisions.

Security contractual requirements (i.e. obligations) placed on a vendor could result in an increase in costs. Requirements for technical control measures need to be identified as early as possible to minimize the costs incurred for implementation.

Procurement of security certified components that meet international standards could assist in reducing the risk of compromise to digital assets. Examples of standards for products and components are Refs [25–27]. In addition, organizations might want to contractually compel vendors to add computer security measures, for instance in the management of computer security risks relating to medical devices[10] used as part of medical treatment programmes.

4.3.5 Organizational and security culture

Given that operating organizations are focused on their primary missions (e.g. health care, industrial radiography, food and agriculture), they might lack an appropriate nuclear and computer security culture, which ultimately has an effect on personnel's nuclear and computer security awareness. It could also significantly complicate the effective implementation of security measures [8].

Security culture is key to an effective security programme and is necessary to ensure that priority is given to information and computer security. Given the nature of cyber-attacks, worldwide connectivity and the unobservable effects of any compromises to digital assets, it is essential that a robust security culture be in place to defend an organization's sensitive digital assets (see Ref. [8]).

5. PROTECTION OF SENSITIVE DIGITAL ASSETS FOR RADIOACTIVE MATERIAL AND ASSOCIATED FACILITIES

Paragraph 4.10 of IAEA Nuclear Security Series No. 13, Nuclear Security Recommendations on Physical Protection of Nuclear Material and Nuclear Facilities (INFCIRC/225/Revision 5) states [28]:

> "Computer based systems used for physical protection, nuclear safety, and nuclear material accountancy and control should be protected against compromise (e.g. cyber attack, manipulation or falsification) consistent with the *threat assessment* or *design basis threat*."

Computer security issues in relation to radioactive material and associated facilities include cyber-attacks that could:

— Degrade physical protection systems resulting in a criminal or other intentional unauthorized act (e.g. blended attacks);
— Manipulate computer based control systems or devices for criminal purposes (e.g. changing the dose for a patient receiving radiation therapy or changing the dose applied to harmful insects or food being irradiated).

Physical protection systems increasingly rely on computer based systems connected to facility networks. Paragraph 3.20 of Ref. [15] states (footnote omitted):

> "States should designate the types of sensitive information that are of security concern and should be protected. These types of information may include:

[10] For further information, see: www.fda.gov/regulatory-information/search-fda-guidance-documents/postmarket-management-cybersecurity-medical-devices

— Details of all computer based systems, including communication systems and instrumentation and control systems that process, handle, store and/or transmit information that is directly or indirectly important to safety or security."

Paragraph 3.21 of Ref. [15] states that "Information security refers to the system, programme or set of rules in place to ensure the confidentiality, integrity and availability of information in any form." Computer security protection for radioactive material and associated facilities is therefore to be integrated into the overall nuclear security programme in order to protect against threats.

5.1. EXAMPLE OF A COMMON INFORMATION SYSTEM ARCHITECTURE

Digital technologies (e.g. IT and OT) provide key services and support to achieve business and security objectives. From a nuclear security perspective, vulnerabilities in these technologies provide opportunities for an adversary to gain access to the organization's networks and potentially exploit this access in order to compromise the functions provided by sensitive digital assets.

The adoption and use of digital technologies have led to the implementation of architectures that prioritize usability and accessibility, with many management functions performed in an automated manner. In many cases, organizations have implemented networks with a single security level (i.e. a flat network). A better option is to implement a segmented network in which key functions are separated into individual levels and zones on the basis of their function, location, consequence or type of communication. Such a segmented architecture would depend on the effective management of network infrastructure and equipment, services and security technologies (e.g. through firewalls or virtual local area networks), and would be supported by an organizational policy and programme that provides holistic security on the basis of defence in depth.

Figure 2 provides a typical model of computer security levels and zones applied to IT and OT networks within a hospital type architecture. The figure shows simple networks and includes computers, switches, firewalls and users. The model uses the following security levels:

— Internet: External users, networks and devices that are not under the control of the facility.
— De-militarized zone (DMZ): A network between the public Internet and the internal networks. The DMZ provides services to both networks.
— IT network: A network that contains the companies' business data and processing, and allows company employees to access resources (e.g. printers, applications, file storage, databases).
— Physical protection system (PPS) network: A network that contains sensitive digital assets and provides protection against criminal or other intentional unauthorized acts that could lead to unauthorized removal or sabotage.
— OT network: A stand-alone device or network that contains sensitive digital assets to support various organizational activities that use radioactive material.
— Sensitive information network: A network or stand-alone device (e.g. a laptop) that processes and stores sensitive information (e.g. detailed elements of a security plan that reveal vulnerabilities and weaknesses in the protection of high activity sources). This network is often in a physically separate location and operated in isolation (i.e. disconnected) from other networks. Any data movement from this network is to be very strictly controlled.
— Security levels: Low, moderate and high levels are used to define the level of requirements and protection necessary on the basis of the risk and consequences of compromise.

In Fig. 2, an adversary who is persistent in the IT network would have to compromise a trusted system or firewall to access the PPS network or the OT network through the IT network, which underlines the advantages of implementing a defensive computer security architecture.

Network firewalls are hardware and software devices that monitor and control traffic between two or more networks (i.e. typically a trusted internal network and an untrusted external network) and control the information flow from one network to the other. A host based firewall is software that operates on the host computer to control network traffic in and out of that computer.

The DMZ is used as a buffer zone between trusted internal networks and publicly accessible systems that are not trusted, such as the Internet. The DMZ can include servers that are public facing (e.g. external website, email servers), which necessitate passing through network firewalls — or other computer security measures — to limit access to specific information on the servers to the trusted internal network only.

A virtual private network (VPN) consists of a server and client component. It allows the user and the server to communicate authenticated data through encrypted channels on public networks. A VPN uses a level of encryption that meets the obligations of the highest security level to ensure that the confidentiality of the communication between the servers (or server) and the client computer is maintained.

Antivirus software, running on compatible computers, monitors the system (i.e. primarily the storage) for signatures that could indicate that a file includes a previously known malware. The user is notified, and the file is either quarantined or deleted from the system to minimize the potential of compromise. This software needs to be regularly updated in order to integrate new malware signatures.

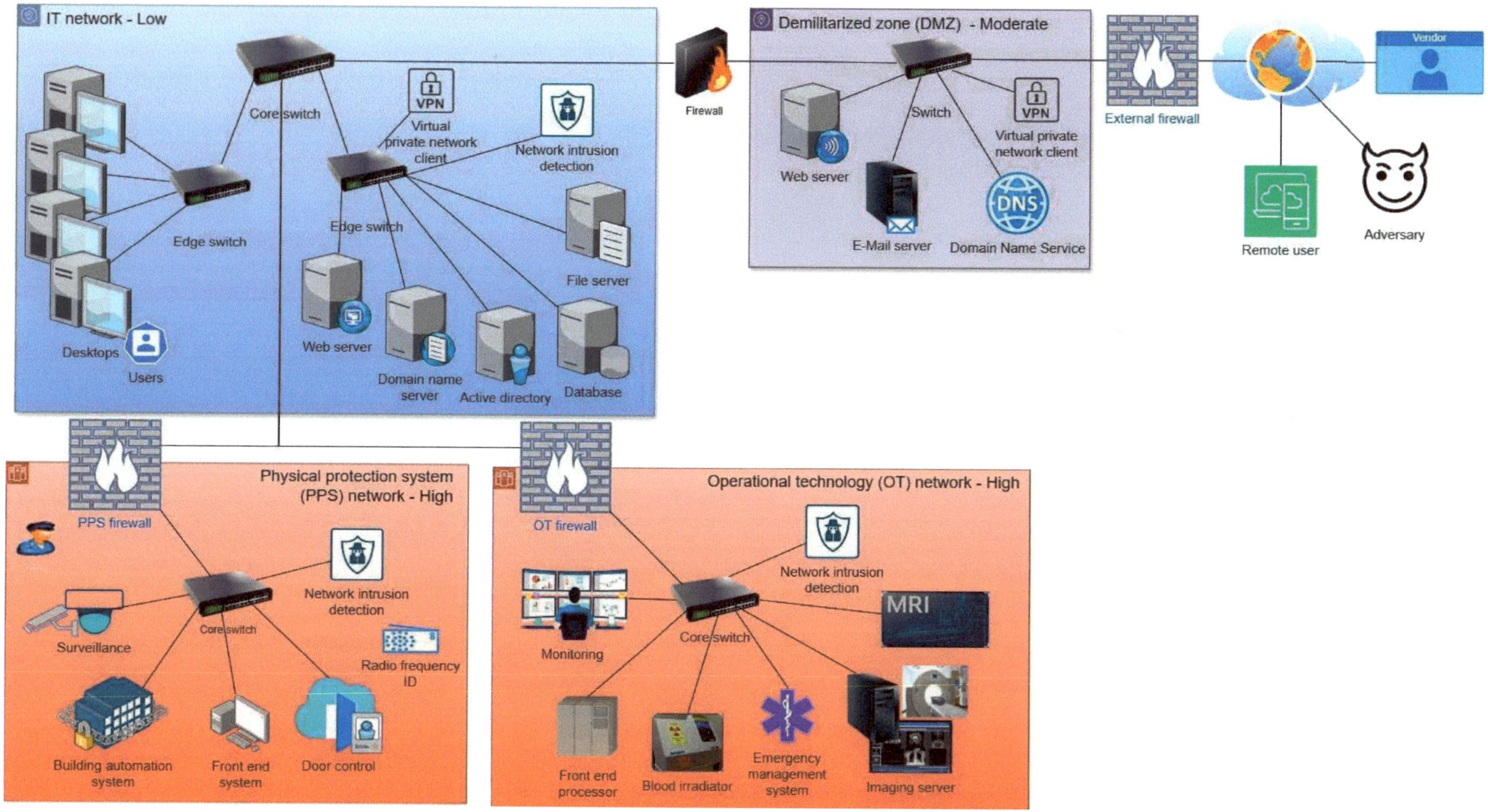

FIG. 2. Simplified IT and OT networks

Intrusion detection for hosting services or networks initiates searches for indications that could identify a cyber-attack and reports when such a detection is made. A system with a prevention capability (e.g. intrusion prevention) could also act on the basis of the intrusion (e.g. blocking the attacking machine or quarantining the compromised machine).

A security information and event management system is a monitoring and alerting software that performs real time analysis on the basis of security logs and incidents generated by applications, hosting services and network devices for early detection and response to abnormal activities.

5.2. OPERATIONAL TECHNOLOGY

OT is a generic term describing analogue devices, digital devices or computers that monitor, manage, control, manipulate or influence real world physical processes. It is an umbrella term that also covers supervisory control and data acquisition systems and distributed control systems, as well as instrumentation and control, industrial control systems and the physical protection system. OT systems typically have components that are inputs (i.e. sensors), outputs (i.e. control elements), controllers and a human–machine interface.

IT and OT have distinct attributes that need to be considered when developing a computer security programme. Table 2 summarizes some of these attributes.

TABLE 2. COMPARISION OF INFORMATION TECHNOLOGY AND OPERATIONAL TECHNOLOGY

Topic	IT	OT
Availability	Delays accepted	Every day, 24 hours a day
Time critical content	Delays are generally accepted	Time critical
Equipment life cycle	Two to three years	5–20 years
Computer security upgrades	Regular and scheduled	Not common practice
Computer security awareness	Good with public resources	Poor
Computer security measures	Very common, easily deployed and updated	Uncommon, and can be difficult to deploy in the case of some devices
Computer security incident response procedures	Well defined and deployed	Uncommon
Security testing and audits	Scheduled and practised	Not well established

5.3. COMPUTER SECURITY DESIGN

Organizations need to understand the design of the systems performing significant functions within their facilities or in the context of their activities. By understanding the design, more effective measures for protection can be developed and implemented in order to minimize the potential for compromise to digital assets.

Legacy operational technology, which includes physical protection systems, were designed using analogue technology and point-to-point communication interfaces, often without computer security in mind. These systems were typically stand-alone, isolated and dedicated to their functions. Physical access was necessary only to interact with the equipment. However, integration of digital technologies, microprocessors and network based communication, including commercial off the shelf technologies, into these systems would result in increased capabilities.

The technological evolution of the PPS and the decreasing availability of analogue components has motivated regulators and operators to analyse the effects of these changes in technology, as well as their impacts on security. The use of digital network based devices increases complexity and vulnerability to attack (i.e. the attack surface[11]). These devices are considered vulnerable to a potential adversary who has access to a device through one of the following attack pathways: network access (e.g. wired, wireless), physical access, access to removable media and portable devices, and supply chain access. These attack pathways could also be exploited through third parties that have privileged access from remote locations.

Section 4 of Ref. [30] states:

> "Supply chain relationships typically involve multiple tiers of globally scattered suppliers throughout the supply chain. This network includes designers, developers, contractors, manufacturers, integrators, solution providers and logistics providers (including shippers, carriers and customs agents). The complexity of these networks provides numerous possibilities for an adversary to compromise a service or device, or to acquire system information prior to use within a facility or organization within the nuclear security regime."

Adversaries are continually developing new tactics, techniques and procedures to target digital technologies, which means that organizations need to employ supplementary computer security measures to ensure the ongoing protection of sensitive information and sensitive digital assets. Not responding to this evolving threat landscape will reduce the ability of an organization to provide detection, delay and response to nuclear security events.

Organizations need to consider the impact of any compromise to PPS functions. Examples of types of compromise that PPS functions could experience include the following [10]:

- The performance of the function is indeterminate; for example, in the case that the access control system is taken over by the adversary, but it does not 'fail', which allows the function to be circumvented on demand.
- The performance of the function demonstrates unexpected behaviour; for example, the camera image does not update, but the clock is running, without any report of failure.
- The performance of the function fails; for example, the video recorder fails to record the camera streams.
- The performance of the function is resilient; no effect can be seen on the system.

5.3.1 Information and computer security requirements

Consistent with the application of a graded approach, design considerations for sensitive digital assets need to determine the degree of protection requirements of such assets on the basis of the following two factors:

(1) The sensitive information that the sensitive digital assets store, process, handle, manage or transmit;
(2) The significance of the functions performed.

To define the level of protection, organizations need to understand the sources of vulnerabilities and the process that adversaries will follow to compromise devices, as well as how impacts on confidentiality, integrity and availability (also known as the 'CIA triad') can degrade system functions and enable sabotage or unauthorized removal of radioactive material.

It is important to understand the process of how an adversary can have an impact on system functions. The process can be described in a simplified manner as follows:

(1) An adversary targets a digital asset to compromise;
(2) The adversary exploits a known or unknown vulnerability (zero day);

[11] The attack surface is the set of points on the boundary of a system, a system component or a digital environment where an adversary can try to enter, have an effect on, or extract data from, that system, component or digital environment (see Ref. [29]).

(3) Step 2 results in an impact on the CIA triad for the exploited equipment, which can be quantified using a common vulnerability scoring system;
(4) The adversary leverages this compromise to have an impact on system functions (e.g. indeterminate, unexpected behaviour, failure).

Understanding the above process allows organizations to effectively define levels of protection and implement computer security measures on the basis of a graded approach (i.e. security levels) so as to ensure the prevention, detection and delay of, as well as the response to, nuclear security threats.

5.3.2 Design of the physical protection system

Reference [1] defines a physical protection system as "An integrated set of physical protection measures intended to prevent the completion of a malicious act." Physical protection systems are a regulatory requirement, and their function is to protect radioactive material. Paragraph 4.11 of Ref. [15] states:

> "The operator should design the security system to deter adversaries from attempting a malicious act and to prevent them from completing such a malicious act through the implementation of detection, delay and response measures. The security system should also include security management measures for the integration of people, procedures and equipment through the application of administrative measures."

Design considerations for information and computer security are to be an integral component of the physical protection system where people, procedures and equipment depend on sensitive information or digital assets.

The first defence against an adversary is detection. Paragraph 4.12 of Ref. [15] states:

> "The operator should implement security measures so that an adversary would encounter detection measures prior to encountering delay measures. The intent of delay measures is to provide response personnel with sufficient time to deploy and interrupt the adversary's efforts to complete a malicious act."

Detection is typically supported by digital assets, which can also be compromised. These digital assets also need protective measures to ensure that the detection function operates as designed to maintain the integrity of alerts.

Once an alert is received by the organization, it needs to be assessed by the appropriate personnel to determine the proper response. Paragraph 4.13 of Ref. [15] states:

> "Most means of detection provide an indirect indication of a potential malicious act. Therefore, when an alarm or other indirect indication triggers that a malicious act might be underway, an assessment should be undertaken to determine its cause. There is always some uncertainty as to the cause of alarms. Alarm assessment requires human observation and judgement, through deployment of response personnel to investigate the cause of the alarm or through use of remote video systems."

Figure 3 provides an example of detection devices and the functions they perform, including data communication, which is processed to generate alerts (i.e. data presentation) to the operators so that an alarm assessment can be performed. Given all of the elements involved in generating an alert, it is important that the proper protection of sensitive digital assets that are part of the detection system are evaluated and assessed in order to maintain the integrity of alerts.

The alarm system's detection devices (e.g. video surveillance, access controls, motion detection) use commercial off the shelf technologies to simplify deployment with standard hardware and software, using wired or wireless networks. Using such technologies, however, increases the susceptibility of the system, and the potential for adversaries with the necessary knowledge to target these digital assets for criminal or other intentional unauthorized purposes using open source or other toolsets. Compromise of these systems could have an adverse impact on the reporting of alarms to the operator or response personnel, generate false alarms or otherwise undermine the ability to provide timely assessments.

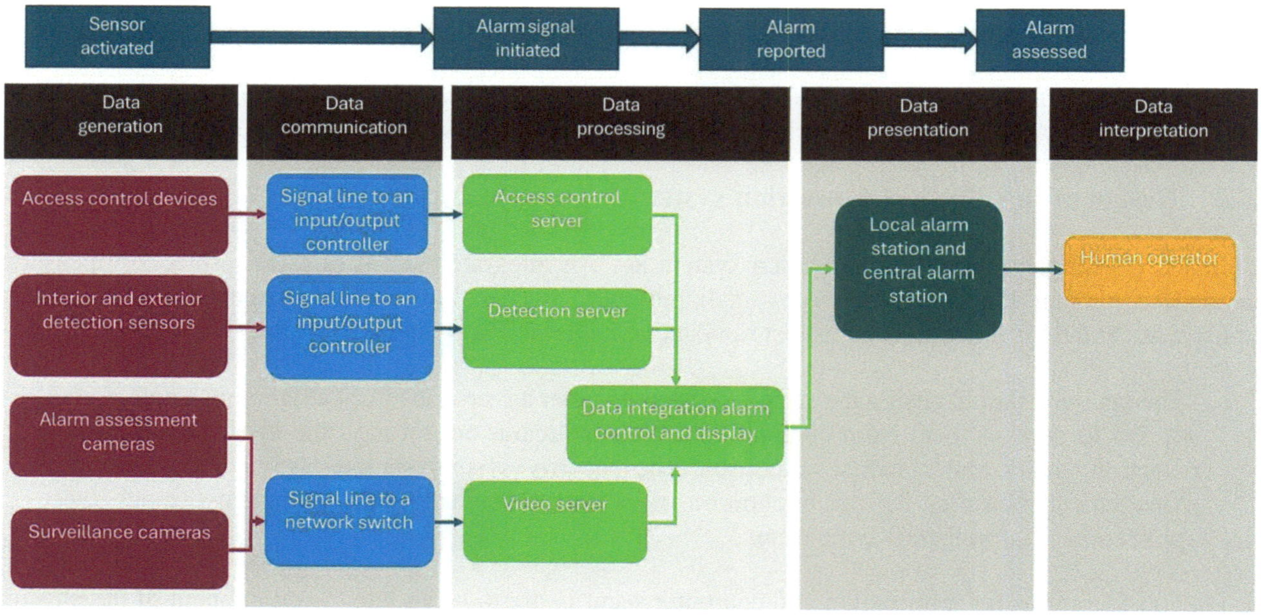

FIG. 3 Alarm detection and assessment.

Paragraph 4.14 of Ref. [15] states that "Assessment activities should be undertaken soon enough after detection and performed quickly enough to enable response personnel to interrupt the adversary prior to completion of the unauthorized removal or sabotage."

A cyber-attack could delay or disrupt the timely response of security personnel. Two examples of cases in which this could happen are the following:

(1) An attack of an alarm system prevents alarms from reaching the central alarm station, which impedes assessment and response measures;
(2) An attack generates false alarms to divert the attention of the response forces to other areas of the facility, to fatigue the response forces, or to ensure that the response forces ignore alarms.

Analysis of historical circumstances surrounding nuclear security events has found that adversaries commonly maintain a presence on the digital assets of an organization for weeks or months prior to detection, initiation of an attack against sensitive data assets, or realization of a criminal or other intentional unauthorized act.

Effective response capabilities involve a sufficient number of personnel with the necessary knowledge, skills (i.e. acquired through training) and tools to respond to incidents in order to defeat adversaries and prevent them from completing their goals.

The organization needs to document the response arrangements for computer security incidents that could escalate to the point of inhibiting the ability of sensitive data assets to perform their functions, thus leading to a nuclear security event. Reference [12] documents response arrangements, which could include the following:

— Methods of ensuring the minimization of impacts to systems supporting safety, security and other response arrangements;
— Defined roles and responsibilities of personnel during nuclear security events;
— Defined communication methods to be used, including between internal (e.g. local physical security) and external stakeholders (e.g. computer security emergency response team, national computer security centre);

— Defined escalation levels and criteria, and methods of interfacing with other security domains;
— Procedures for reporting security events, including any reporting obligations and arrangements for review of the physical protection system after a nuclear security event, and any corrective actions necessary.

Specialized computer security expert resources will generally be necessary for computer security response activities, and these resources will likely need to be acquired externally (i.e. provided by an outside organization other than the operator). Such resources can typically be provided through private computer security organizations, national technical support organizations or through international assistance. When accessing these expert resources, planning, communication and coordination are essential to ensure that the response is timely and effective.

In terms of incident response, the physical protection system incorporates computer security measures that assist in the detection of, and response to, any compromise of digital assets. A focus on timely detection will ensure that the effects of compromise can be mitigated, or a response can be formulated, before the realization of a nuclear security event. Paragraph 4.16 of Ref. [15] states that:

> "The security system should be designed to provide adequate protection against all defined threats along all possible adversary pathways to the target. Along any possible pathway, detection measures, delay times and resulting responses should combine to protect the target."

Computer security is necessary to support the overall security of the facility. Paragraph 4.17 of Ref. [15] states that:

> "A security system should employ the principle of defence in depth, such that several layers and methods of protection (structural, technical, personnel and organizational) need to be overcome or circumvented by an adversary in order to achieve his or her objective."

Defence in depth for computer security is provided by the arrangement of digital assets into computer security levels and zones that have hardened boundaries (physical and logical). The resulting defensive computer security architecture provides protection by ensuring that the levels and zones are arranged on the basis of the information and data flows between digital assets at the facility [10].

5.4. PHYSICAL PROTECTION SYSTEMS

The functions of physical protection systems could include the following:

— Access control of personnel to controlled areas;
— Physical intrusion detection and protection of material;
— Video surveillance of controlled areas to provide alarm assessment (e.g. provisions to detect unauthorized intrusions);
— Alarms for identified nuclear security events (e.g. unauthorized access to secured areas, radiation levels, device tampering);
— Prevention of the unauthorized removal of, or of access to, material (e.g. tamper indicating devices);
— Notification of the status of physical security to on-site and/or off-site response forces;
— Monitoring of area radiation fields to detect tampering or the unauthorized removal of radioactive sources.

Having a combination of diverse and physical separated areas (e.g. public, protected, locations with radioactive material) within the design of the physical protection system, with the associated networks and devices, makes it difficult to segregate levels of protection on the basis of consequences [10]. For example, a physical protection system supports the protection of the entire facility, with the devices and networks spanning multiple areas (e.g. secure and publicly accessible areas), and cannot be separated into different security levels if all of the functions need to be supported by all of the sensitive data assets. An evaluation needs to be performed to consider whether the greater security benefit provided through the functions supporting this interconnectedness outweigh the increase in the susceptibility to potential cyber-attacks.

Designing a defensive computer security architecture will entail the implementation of a segmented physical protection system network, with each segment being assigned one or more significant security functions. It is key to address the integration of these functions in a manner that maintains separation of the sensitive data assets, with their potential to act as pathways for compromise, in order to ensure defence in depth. For example, without a thorough application of a defensive computer security architecture, a wireless camera for perimeter monitoring could be used by a competent adversary to achieve direct remote access to the entry control system database simply by compromising the camera.

5.4.1 Design principles of the physical protection system

IAEA Nuclear Security Series No. 40-T, Handbook on the Design of Physical Protection Systems for Nuclear Material and Nuclear Facilities [31], identifies physical protection system networks that have different designs in accordance with the size of the system and the regulatory requirements for the organization and the facilities that the physical protection system supports. Many variables are to be considered when designing a network that will enable flexibility and growth. Some of the most important engineering principles include the following:

— Hierarchy;
— Modularity;
— Confidentiality;
— Availability (or resiliency);
— Flexibility;
— Network integrity;
— Complexity.

Each of these engineering principles is fully described in para. 6.6 of Ref. [31].

5.4.2 Access control systems

Access control systems, such as keys and locks, combination locks, badges or readers and electronic locks are used to allow personnel access to secured areas while restricting the access of unauthorized persons.

Information and computer security concerns for access control systems include the following:

— The confidentiality of personally identifiable information (e.g. biometrics, names, personal identification numbers, employee identification information);
— The integrity of access control lists, authentication, control logic;
— The availability of access control system devices.

5.4.3 Physical intrusion detection systems

Physical intrusion detection systems are intended to alert facility personnel of an unauthorized breach. These intrusion detection systems use electronic alarm sensors, video surveillance and other means to detect such breaches.

Information and computer security concerns in relation to the detection of physical intrusions include the following:

— The confidentiality of information that provides details concerning the types and locations of the sensors;
— The integrity of the alarm networks and logic;

— The availability of the physical intrusion detection networks to transmit alarms to the alarm monitoring stations.

5.4.4 Video surveillance systems

Video surveillance is primarily intended to enable remote assessment of security alarms. Video signals are transmitted, typically over a network, to an alarm monitoring station where an operator can view and properly assess the cause of the alarm. Information and computer security concerns for video surveillance include the following:

— The confidentiality of the video feed (e.g. whether it can be intercepted and viewed);
— The integrity (alteration) of the camera video feed;
— The availability of the cameras or their networks.

5.5. OPERATIONAL TECHNOLOGIES USED IN RADIOLOGICAL DEVICES

Radiological devices (e.g. medical devices, food irradiators, industrial irradiators) perform functions that offer an important social benefit. These devices are usually subject to equipment maintenance, although the application software and the operating system are rarely updated or patched after installation.

Information and computer security concerns[12] for radiological devices could include the following:

— Changes in operation or sabotage of a device, resulting in incorrect radiation exposure (e.g. overdose, underdose, wrong treatment location) for the patient or organization personnel;
— Modification of the operation to allow for easier removal of the device or the radioactive material within the device;
— Ransomware, for example the encryption of the system or data to prevent use of the system (i.e. loss of availability);
— Loss or disclosure of personally identifiable information or patient data;
— Unauthorized access to devices through wireless networks.

5.6. IMPLEMENTATION OF COMPUTER SECURITY

Security policies and procedures (e.g. concerning computer security awareness, removable media, incident response, remote access) need to be established and supported through education, training and assurance activities (e.g. exercises, drills, assessments), and reinforced by a security culture that promotes continual improvement and progression of maturity in organizational processes.

Computer security also involves ongoing activities to install and test firmware/software updates, change or remove default passwords and harden systems from their default configuration (e.g. removing unnecessary services and applications). In some cases, and particularly with the firmware and software for operational technology and the physical protection system, updates will not be available. It is important that vulnerabilities are assessed, and compensatory measures established to prevent exploitation (i.e. the best case) or to provide immediate detection of exploitation (i.e. the worst case).

System hardening provides a long term benefit in that it decreases the attack surface of a system or device. However, it is possible that any commercial off the shelf equipment being used does not allow for the configuration of hardening, either by design or contractually (e.g. information is not provided by the vendor).

---

[12] While some of the concerns presented in the list are not directly connected with nuclear security, they have been included because they are taken into consideration during the development of the overall computer security programme. Computer security for patient safety could also be covered under the State's healthcare regulations relating to medical devices.

In such cases, additional technical controls will need to be introduced to minimize access to attack pathways (e.g. wired networks, wireless networks, physical access, portable media, mobile devices, supply chain access).

6. IMPLEMENTATION OF AN INFORMATION AND COMPUTER SECURITY PROGRAMME FOR RADIOACTIVE MATERIAL

It is important for an organization to define its objectives, demands, processes and means of managing information and computer security, typically implemented through an information and computer security programme. The information and computer security programme could be a stand-alone or separate programme, or it could be part of the overall security plan.

The information and computer security programme reflects the resources needed to implement the programme after taking into consideration information and computer security demands. Demands associated with the computer security programme are typically managed internally, or by awarding contracts in the case of a lack of internal expertise. It is important that the operator allocate an adequate budget to guarantee that the necessary computer security activities are performed.

The list below is an example of the contents of a computer security programme that is integrated into the security plan. This general case is detailed in Ref. [5] as the minimum contents of a computer security programme and include the following:

— Organization and responsibilities;
— Risk, vulnerability and compliance assessment;
— Digital asset management;
— Computer security design and architecture;
— Operational security procedures;
— Personnel management.

Subsections 6.1–6.6 examine each of the above areas in more detail, analysing the individual elements comprising these areas, examining their current status and providing suggestions for changes to the security plan, the rationale for changes, and examples.

6.1. ORGANIZATION AND RESPONSIBILITIES

The primary accountability for computer security functions is to be designated to a single qualified individual in the computer security programme, with further responsibilities assigned to other personnel. Where responsibilities are assigned to a vendor, specific contract language needs to document the obligations. Entities having responsibilities for computer security functions are to be trained appropriately and need to understand their duties as they relate to the security of radioactive material.

(1) Organizational charts

Status: Exist in the security plan but additional provisions would need to be added for information and computer security.

Suggested changes: Update the security plan.

Rationale: Information and computer security involve specific requirements and activities that could be outlined in the security plan. Roles in relation to computer security might also need to be added to the security plan, as well as information concerning specific expertise, which will likely be provided by external contract personnel, by a third party or by a vendor.

(2) Responsible persons and reporting responsibilities

Status: Information is already included in the security plan, but additional provisions would be needed for information and computer security.

Suggested changes: A person will need to be assigned the role and responsibility of providing oversight for information and computer security. Senior management will need to ensure sponsorship and promotion through adoption of an information security policy, and creation or modification of a job role to cover information and computer security.

Rationale: A single person or entity needs to have ownership of information and computer security to ensure that related activities are broadly and consistently applied. Senior management support is crucial in ensuring that the appropriate priority is given to these activities.

(3) Interfaces with other security and support organizations

Status: Such interfaces already exist in the security plan for other subject areas, but additional provisions would be needed for information and computer security.

Suggested changes: The security plan will formalize interfaces between computer security and other security and support organizations, including physical interfaces, along with those for information and security personnel, and for IT and network support organizations.

Rationale: Effective computer security depends on other security disciplines, and the work of network and IT support organizations.

(4) Enforcement and corrective actions

Status: Such actions exist in the security plan for other subject areas, but additional provisions would be needed for information and computer security.

Suggested changes: New or updated policies need to be implemented to cover sensitive information, user access control, the password policy and the acceptable use policy to outline potential violations with the associated penalties. Such policies could also be included in the employment contracts for personnel or in the conditions of employment. In addition, the corrective action programme and incident response mechanism could be aligned with IT activities (e.g. help desk, configuration updates, monitoring, login banners).

Rationale: Cyber-attacks could lead to the theft of intellectual property, sensitive information or system design details that include vulnerabilities. As a result, adversaries could plan physical attacks that have an increased likelihood of success. It is particularly difficult to thwart insider adversaries with cyber skills, but detection activities can act as a significant deterrent.

(5) Periodic review and approval process

Status: The periodic review and approval process is already covered in the security plan.

Suggested changes: Review the security plan to ensure that the review and approval process can apply to information and computer security.

Rationale: While information and computer security add specific demands and activities to support security, these are not expected to result in major changes to the review and approval process of the organization, nor to responsibilities.

6.2. RISK, VULNERABILITY AND COMPLIANCE ASSESSMENT

Given the evolving threat landscape, organizations need to conduct continual risk, vulnerability and compliance assessments. These assessments involve thorough testing, audits and reassessments to identify and mitigate risks associated with sensitive digital assets relating to radioactive material. In addition, assessments will closely align with evolving regulatory frameworks, adapting to changes in legislation in order to maintain compliance and to enhance the overall security posture of radioactive material.

(1) Periodic review and reassessment of the computer security programme

Status: Such a review and reassessment process exists in the security plan, but additional provisions are needed for information and computer security.

Suggested changes: Establish an information and computer security programme within the security plan. Integrate and leverage the current process to review and reassess the security plan for the computer security programme. Additional resources to cover information and computer security could be necessary for the review.

Rationale: The information and computer security programme represents the objectives, policies, processes and criteria for success (i.e. metrics, observables) against which the success or deficiencies of the programme are assessed. Additionally, given the specialized demands of information and computer security, a diverse group of experts is necessary to ensure the effective results of the assessment.

(2) Self-assessment (including active and passive testing procedures)

Status: Such an assessment already exists in the security plan, but additional provisions are needed for information and computer security.

Suggested changes: Establish standards and procedures on how to self-assess the equipment and systems that are an integral part of information and computer security, along with measurement criteria and expected results.

Rationale: Standards and procedures are necessary to set a baseline against which progress can be measured. This is an important part of any continuous improvement process. The process needs to be repeatable and lead to consistent, high quality results. Qualified personnel or external contractors will need to have the necessary capabilities and expertise to ensure the efficiency and validity of such assessments.

(3) Periodic and reactive risk reassessments and associated methodologies

Status: Such reassessments and methodologies exist in the security plan, but additional provisions are needed for information and computer security.

Suggested changes: Implement a risk assessment methodology (see Refs [10, 21]), which includes risk identification, risk analysis and risk evaluation, as well as communication and consultation, monitoring and review. Review of other organizations' activities (e.g. through enterprise risk management, business continuity) to highlight the need to report relevant details concerning information and computer security, since these details could underline increased or new risks.

Rationale: Risk management is central to nuclear security and consists of a robust and repeatable process that is overarching. Information and computer security intersects with various other areas. For example, cyber-attacks on the IT networks could migrate to the PPS or OT networks, causing significant disruption or a potential compromise to nuclear security.

(4) Audit procedures and records to correct deficiencies

Status: Such procedures exist in the security plan, but additional provisions are needed for information and computer security.

Suggested changes: Procedures could be reviewed and modified to include information and computer security and to address deficiencies in this regard.

Rationale: Procedures and tracking lead to records, which can be reviewed during a self-assessment and audit process in order to identify areas for improvement. For new programmes, it is important to generate records in a consistent manner so as to ensure that significant data is identified and leads to the effective discovery and correction of deficiencies.

(5) Review of legislative and regulatory compliance

Status: Such reviews exist in the security plan, but additional provisions are needed for information and computer security.

Suggested changes: Review and consult with experts on regulations and legal considerations to ensure that all legal and regulatory requirements are met for information and computer security. Some of the key computer security regulatory elements that are considered in the computer security programme include the following:

- Identification of functions, systems and sensitive digital assets;
- Risk management;
- Protection of a system function against compromise;
- Configuration and change management;
- Supply chain management;
- Incident response.

Rationale: Many States have enacted laws and regulations concerning information and computer security, along with mandatory cyber-attack reporting requirements. It is crucial for organizations to understand the compliance requirements and how the organizations are to meet them. Computer security activities can be performed to meet a number of legal and regulatory requirements.

6.3. DIGITAL ASSET MANAGEMENT

Digital asset management is essential for ensuring the protection of radioactive material. A comprehensive inventory of all digital assets is to be maintained, with sensitive digital assets identified on the basis of the consequences if compromised, in order to prioritize assets and determine their significance (e.g. contribution to security, safety and sensitive information management functions). Sensitive digital assets are then properly categorized at the appropriate level of security to ensure compliance with regulatory requirements.

(1) List all of the digital assets

Status: This is a new item, since it does not exist in the security plan. It can, however, be common practice in other organizations (IT departments), although the information could be incomplete in the case of nuclear security operational environments.

Suggested changes: Create a comprehensive list of digital assets, their functions and their locations.

Rationale: Understanding core functions and supporting computer equipment is the first step in a computer security programme, which includes identifying all digital assets that will need to be holistically managed.

(2) Identify and categorize sensitive digital assets

Status: This item is new, and not included in the security plan. A new process is therefore necessary.

Suggested changes: Create a procedure for owners of sensitive digital assets to classify or categorize the assets, which results in an association to a security level and the appropriate security requirements.

(3) Create a comprehensive list of sensitive data assets, their functions and the sensitivity of the information, their location and the associated consequences of compromise.

Rationale: States are encouraged to have additional regulatory requirements for sensitive digital assets (see Ref. [9]). The list of sensitive digital assets will need to be complete and verified on a regular basis so as to ensure that the risk to such assets is managed in a holistic manner. It is a crucial step in risk evaluation to determine the magnitude of the risk on the basis of the worst case effects of compromise to the sensitive data assets. The risk treatment options, along with communication and consultation steps, are part of this evaluation, which includes the assignment of a security level to the sensitive data assets.

(4) Identify and document all data flow and network diagrams

Status: This item is new, since it is not included in the security plan.

Suggested changes: Create a comprehensive document for sensitive digital assets, which details data flow and network diagrams, including all connections to internal and external computer based systems.

Rationale: Cyber-attacks are likely to be network based attacks and can result in the undetected compromise of one or more sensitive digital assets. An understanding of information flows and network diagrams aids in the design and implementation of computer security measures, such as firewalls, removable media controls, physical protection measures and incident response.

(5) Manage configurations of sensitive digital assets

Status: This item exists in the security plan (or the IT security plan), but many additional provisions are necessary for information and computer security.

Suggested changes: Create a comprehensive document that lists the configurations of sensitive digital assets (e.g. hardware, firmware, software applications, equipment status and associated configurations).

Rationale: It is important to understand the configurations of sensitive digital assets, and in particular those that could lead to vulnerabilities, weaknesses or the existence of possible attack vectors that could result in the compromise of such assets. The above document will need to be completed and then verified on a regular basis to ensure that vulnerabilities and the configurations of sensitive digital assets are effectively managed to support nuclear security. Detailed configuration information can also provide support for detection (e.g. unauthorized software or data) and recovery activities.

6.4. COMPUTER SECURITY DESIGN AND ARCHITECTURE

A defensive computer security architecture provides protection mechanisms for important functions and sensitive digital assets through an architectural design that encompasses multiple levels of security requirements, spanning the life cycle of the organization so as to ensure a resilient architecture.

(1) Define the architecture and design principles

Status: The item is not always outlined in the security plan, or it is presented in a non-exhaustive manner, meaning that adaptation is necessary.

Suggested changes: Establish a set of objectives that ensure adequate decoupling of security levels and zones within a defensive architecture. The objectives of the architecture are to consider the overall security objectives of the organization.

Rationale: Implementing security level requirements that include a priority to ensure that strategic objectives are met. Architectural requirements are the highest priority to increase the difficulty of an adversary to successfully disable, delay or degrade security functions or response measures.

(2) Implement fundamental security design approaches

Status: Fundamental security design approaches are not always outlined in the security plan or are presented in a non-exhaustive manner. Adaptation is therefore necessary.

Suggested changes: Define and use security levels and zones that exist within the organization. The security levels are also to be accompanied by a list of demands that correspond to the function, deployment and use of the sensitive data asset.

Formalize the interface with the IT and PPS departments to ensure that the sensitive data assets are added to the appropriate levels and zones, which meet both the architectural objectives and the security level objectives.

Rationale: An organizational understanding of security objectives and how they are addressed will result in more secure designs that are compatible with the overall security strategy of the organization.

(3) Implement computer security requirements for vendors, contractors and suppliers

Status: This item is not always included in the security plan, or is presented in a non-exhaustive manner. Adaptation is therefore necessary.

Suggested changes: Modification of procedures or standards for procurement, development and acquisition of services and equipment.

Rationale: The identification of such procedures or standards early in the development or acquisition stages has the potential to reduce the costs of both implementation and deployment, while providing further options and alternates for computer security measures to fulfil objectives.

(4) Ensure security throughout the life cycle

Status: Full life cycle security exists in the security plan, but additional specifics for information and computer security could be relevant.

Suggested changes: Create processes to determine information and computer security demands for the life cycle stages of sensitive digital assets that are applicable to the organization, namely acquisition, testing, installation, commissioning (i.e. deployment), operations and maintenance, and disposal (i.e. decommissioning).

Rationale: Sensitive digital assets will have different security objectives on the basis of the life cycle stage. It is important to identify such objectives — at least on a generic level — to ensure that the security objectives are met throughout the life cycle. For example, a device that is not sufficiently protected during the acquisition stage could be compromised before deployment and has the potential to cause harm when in operation.

6.5. OPERATIONAL SECURITY PROCEDURES

Operational security procedures are necessary for an information and computer security programme to support the proper implementation and maintenance of access control, data security, communication, monitoring and maintenance that ensures the safe and secure functions of nuclear security, thereby contributing to continuity management. These operational procedures collectively reinforce the resilience and continuity of nuclear operations, underscoring the paramount importance of maintaining a secure and controlled environment at facilities.

(1) Access control

Status: Access control procedures exist in the security plan, but additional specific procedures for information and computer security could also be necessary.

Suggested changes: Review and modify access control procedures to ensure that computer security objectives are included and enforced for access to sensitive data assets and sensitive information.

Rationale: Access to sensitive data assets represents a significant risk to the security of the facility. Privileged accounts, such as system administrators, security personnel, vendors and other third parties, need to be appropriately managed. Control of access needs to consider physical access, network access (e.g. wired, wireless), removable media and portable devices, along with access by third parties or from remote locations. Applying effective computer security measures in relation to the protection and control of sensitive digital assets will likely be one of the most significant controls that operators can employ.

(2) Data security

Status: Data security procedures exist in the security plan, but additional specific procedures for information and computer security might be necessary.

Suggested changes: Review and modify data (and information) security procedures to ensure that security objectives are met and enforced for data and sensitive information that are associated with sensitive data assets.

Rationale: Data exists in many forms. Sensitive data assets are information technology and have the potential to create, process, store, transmit and delete vast quantities of information. Precautions are thus to be taken to protect against either leaks or the theft of potentially sensitive information. Information leaks could include uploading of data to personal devices or to the cloud, posting on social media or open Internet forums, or the loss of digital assets with sensitive information.

(3) Communication security

Status: Communication security exists in the security plan, but additional provisions are necessary for information and computer security.

Suggested changes: Review and modify communication security procedures to ensure that security objectives for networks and information transfer are met.

Rationale: The operational control of networks and other information transfer processes is essential to ensure security in the delivery process. Network segmentation, management and monitoring are key measures that need to be rigorously controlled within operating procedures. These measures will substantively reduce the risks associated with cyber-attacks that could leverage network and information transfer pathways.

(4) Application and platform security

Status: Application and platform security (e.g. hardening, patch management, malware protection) are not always included in the security plan or could be presented in a non-exhaustive manner. Adaptation might therefore be necessary.

Suggested changes: New standards and procedures to ensure that sensitive digital assets are hardened to the greatest extent possible in order to reduce the attack surface and vulnerabilities can largely be managed through patch management and updates, or malware detection and protection.

Rationale: Hardening of sensitive digital assets provides a long lasting and quantifiable reduction in risk, while at the same time limiting additional measures to compensate for, or detect the exploitation of, unpatched vulnerabilities.

(5) System monitoring

Status: System monitoring (including log management) is not always in the security plan or is presented in a non-exhaustive manner. Adaptation could therefore be necessary.

Suggested changes: New standards and procedures are needed to ensure that sensitive data assets provide the capability to generate alarms and system logs, offering relevant computer security information to aid in detection, response and recovery, and also ensuring continual improvements to computer security measures.

Rationale: System monitoring is important to provide early indications of potential computer security incidents that could result in nuclear security events. Moreover, such monitoring deters potential insider adversaries from committing criminal or other intentional unauthorized acts since it is a very effective approach to detecting potential policy or security violations resulting from unauthorized insider activity.

(6) Computer security maintenance

Status: Computer security maintenance is not always included in the security plan or is presented in a non-exhaustive manner. Adaptation could therefore be necessary.

Suggested changes: New standards and procedures to ensure that sensitive data assets, computer security levels and zones, and their respective measures, are maintained so as to remain effective (see Subsection 7.3 on maintenance activities).

Rationale: The effectiveness of any security measure decreases over time and necessitates periodic updates and improvements, as well as the correction of identified deficiencies, either prompt or immediate, depending on the risk.

(7) Incident handling

Status: Incident handling exists in the security plan, but additional provisions are necessary for information and computer security.

Suggested changes: New procedures can be established, along with computer security incident response responsibilities, to create structure, decision making authority, expertise and competence. Personnel need to understand and be capable of defining a computer security incident, as well as identify when such incidents can result in a nuclear security event. Contractual arrangements for third party or vendor support to respond to incidents are also needed. Such considerations can also take into account the safety response.

Rationale: Facilities using radioactive material, as well as associated facilities, organizations and activities, have a responsibility to protect sensitive information and sensitive information assets that, if compromised, could have an effect on nuclear security. This protection includes the development of contingency and response plans, along with the capabilities to address criminal and other intentional unauthorized acts, including cyber-attacks. Effective response to cyber-attacks can significantly limit the severity of the impact and consequences.

(8) Business continuity and disaster recovery

Status: Business continuity and disaster recovery exist in the security plan, but additional provisions are necessary for information and computer security.

Suggested changes: Review and evaluate the maximum tolerable disruption of sensitive data assets, including the associated consequences. A report needs to be made, and if necessary, an update of business continuity and disaster recovery plans performed on the basis of the worst possible cyber-attack scenario, including blended attack scenarios.

Rationale: Cyber-attacks have the potential to disable or degrade the functions performed by multiple systems and/or ensure the denial of access to shared resources. Business continuity and disaster recovery need to provide for the continued ability of the organization to meet its nuclear security and safety requirements.

(9) System backup

Status: System backup exists in the security plan, but additional provisions are necessary for information and computer security.

Suggested changes: Review and revise system backup procedures to protect against the effects of cyber-attacks on the system (e.g. denial of service, destruction, compromise or ransomware) and on its backup media. Such reviews and revisions will ensure that systems can be recovered after eradication of malware from the environment.

Rationale: Nuclear security functions are performed by specialized systems that are under strict configuration control and management. The recovery of these systems depends on regular backups. The availability and integrity of these backups therefore need to be ensured.

6.6. PERSONNEL MANAGEMENT

Personnel management play a key role in upholding the trustworthiness and reliability of individuals entrusted with sensitive information and crucial functions by implementing background checks, behavioural analysis and termination procedures to protect sensitive information, assets and systems. In addition, a computer security awareness and training programme is essential to allow personnel to acquire the knowledge and skills to protect digital assets and prevent unauthorized access. A comprehensive personnel management approach is integral to maintaining the highest standards of nuclear security.

(1) Trustworthiness checks

Status: Trustworthiness checks (i.e. personnel vetting) exist in the security plan, but additional provisions are necessary for information and computer security.

Suggested changes: Review and evaluate personnel access, authorization and privileges in relation to sensitive digital assets. For sensitive operations (e.g. system administration, network monitoring), personnel need to be subject to additional trustworthiness checks. Giving specific personnel multiple privileges for all or for a significant number of sensitive data assets in an effort to provide defence in depth is to be avoided, where possible. If unavoidable, such personnel are to be subject to additional or more rigorous checks.

Rationale: Trustworthiness includes the vetting of third parties, personnel that have access to radioactive material, and personnel that operate or could alter the operation of the sensitive data assets, including the physical protection system. Trustworthiness checks are to be implemented using a graded approach on the basis of whether personnel have access to sensitive information (see Ref. [15]), and of the significance of compromise. Potential insider adversaries could use aggregate or global privileges to compromise multiple sensitive data assets and/or systems, possibly causing severe impacts. Such insiders represent the most capable threats for facilities. Trustworthiness checks, as well as behaviour monitoring and observation, supported by technical control measures can provide effective protection against such threats.

(2) Awareness raising and training

Status: Awareness raising and training are addressed in the security plan, but additional provisions are necessary for information and computer security.

Suggested changes: Review, evaluate and revise the training programme to incorporate information and computer security awareness training.

Rationale: The human factor is a significant aspect of many cyber-attacks. Personnel training and awareness raising activities thus provide preventive measures since they aim to reduce the susceptibility of personnel to social engineering techniques (e.g. phishing, waterhole attacks). Social engineering techniques allow the adversary to coerce personnel into providing credentials, access and privileges to internal networks and resources.

(3) Qualification of personnel

Status: Qualification of personnel is addressed in the security plan, but additional provisions are necessary for information and computer security.

Suggested changes: Review of job roles and descriptions to determine the qualifications (either internal or external) necessary to perform roles that have significance to information and computer security.

Rationale: Having personnel who are qualified to take actions that support computer security increases overall confidence that such actions are being undertaken with a high degree of quality. If the configuration of security appliances by underqualified personnel is performed poorly, for example, it can severely degrade the overall security of the organization.

(4) Termination or transfer of personnel

Status: Measures to terminate or transfer personal are outlined in the security plan, but additional provisions are necessary for information and computer security.

Suggested changes: Review and revise employment and vendor contracts and the code of conduct policy to enable the dismissal or discipline of personnel who violate policies or clauses that are important to information and computer security.

Rationale: These elements provide an overall level of deterrence against insider adversaries, and promote compliance with such policies among all personnel and third parties. The result is likely to be an overall risk reduction in terms of human performance or insider threats.

7. SUSTAINABILITY OF AN ORGANIZATION'S COMPUTER SECURITY PROGRAMME

It is important to conduct effectiveness and sustainability activities so as to maintain and improve the information and computer security programme. Such activities provide oversight and ensure that computer security activities are being performed by personnel and contractors in a way that continues to meet the rigour of the determined level of security, as detailed in the security programme. The operator also needs to maintain up to date threat information so as to adapt the security plan in a way that can address such threats.

7.1. MAINTENANCE OF A COMPUTER SECURITY PROGRAMME

The operator needs to ensure effective maintenance of a computer security programme to maintain systems or components in good operating condition. These maintenance activities need to be analysed periodically in order to determine if changes are needed through a configuration control and change management process for associated administrative, technical and physical control measures. These maintenance activities could include:

— Periodic preventive maintenance or testing, such as backups, software and firmware updates, log reviews and computer security vulnerability assessments;
— Actions to detect, preclude or mitigate the degradation of components;
— Actions to diagnose, repair, overhaul or replace failed components.

Maintenance procedures include documented instructions for computer security measures, such as checks and balances (e.g. a two person rule for configuration changes on operational systems), and ensure that systems are correctly re-established after maintenance [9].

7.2. COMPUTER SECURITY ASSURANCE ACTIVITIES

Computer security exercises and assessments are to be conducted throughout the organization to provide a continuous improvement process that can identify gaps and evaluate the organization's effectiveness in relation to the computer security programme. Assurance activities need to be repeatable and reliable, and are to be conducted on a periodic basis, as well as whenever a computer security incident occurs or the threat assessment changes. The output of such exercises and assessments will include the identification of deficiencies, good practices and suggestions for improvements, and will ultimately result in corrective action plans and mitigation actions to defend against threats.

Reference [12] provides details of assurance activities applicable to nuclear facilities, which can be adapted to other radioactive material and associated facilities. Exercises and testing could be used for other elements of the computer security programme, such as security procedures and personnel management. Response exercises could also be conducted to verify and validate procedures for continuous improvement so as to be prepared for incident response.

Assurance activities could be performed by internal or external groups. Computer security assessment, for example, could be performed by an internal team as a self-assessment activity. If such an assessment is performed by external groups, the results would need to be verified internally.

The trustworthiness of independent or external assessors is to be determined before they are permitted access to the information or to the facility since assurance activities are likely to involve sensitive information in relation to computer security. Further information on trustworthiness checks is provided in Ref. [6]. Appropriate restrictions are to be in place for the removal, use, storage, distribution and destruction of sensitive information.

Capabilities to conduct assurance activities are to be developed and reinforced at regular intervals to keep pace with changes in technology and threats. Such capabilities are needed by both the personnel who perform the assurance activities and the competent authority, who might need to review the results of these activities.

REFERENCES

[1] INTERNATIONAL ATOMIC ENERGY AGENCY, IAEA Nuclear Safety and Security Glossary: Terminology Used in Nuclear Safety, Nuclear Security, Radiation Protection and Emergency Preparedness and Response, IAEA, Vienna (2022 (Interim) Edition), https://doi.org/10.61092/iaea.rrxi-t56z

[2] INTERNATIONAL ATOMIC ENERGY AGENCY, Nuclear Security Recommendations on Radioactive Material and Associated Facilities, IAEA Nuclear Security Series No. 14, IAEA, Vienna (2011).

[3] INTERNATIONAL ATOMIC ENERGY AGENCY, Objective and Essential Elements of a State's Nuclear Security Regime, IAEA Nuclear Security Series No. 20, IAEA, Vienna (2013), https://doi.org/10.61092/iaea.ajrj-ymul

[4] INTERNATIONAL ATOMIC ENERGY AGENCY, Security of Nuclear Information, IAEA Nuclear Security Series No. 23-G, IAEA, Vienna (2015). (a revision of this publication is in preparation.)

[5] INTERNATIONAL ATOMIC ENERGY AGENCY, Computer Security for Nuclear Security, IAEA Nuclear Security Series No. 42-G, IAEA, Vienna (2021).

[6] INTERNATIONAL ATOMIC ENERGY AGENCY, National Nuclear Security Threat Assessment, Design Basis Threats and Representative Threat Statements, IAEA Nuclear Security Series No. 10-G (Rev. 1), IAEA, Vienna (2021).

[7] INTERNATIONAL ATOMIC ENERGY AGENCY, Preventive and Protective Measures against Insider Threats, IAEA Nuclear Security Series No. 8-G (Rev. 1), IAEA, Vienna (2020).

[8] INTERNATIONAL ATOMIC ENERGY AGENCY, Nuclear Security Culture, IAEA Nuclear Security Series No. 7, IAEA, Vienna (2008).

[9] INTERNATIONAL ATOMIC ENERGY AGENCY, Computer Security of Instrumentation and Control Systems at Nuclear Facilities, IAEA Nuclear Security Series No. 33-T, IAEA, Vienna (2018).

[10] INTERNATIONAL ATOMIC ENERGY AGENCY, Computer Security Techniques for Nuclear Facilities, IAEA Nuclear Security Series No. 17-T (Rev. 1), IAEA, Vienna (2021).

[11] INTERNATIONAL ATOMIC ENERGY AGENCY, Nuclear Security Management for Research Reactors and Related Facilities, Non-serial Publications, IAEA, Vienna (2016).

[12] INTERNATIONAL ATOMIC ENERGY AGENCY, Computer Security Incident Response Planning at Nuclear Facilities, Non-serial Publications, IAEA, Vienna (2016).

[13] INTERNATIONAL ATOMIC ENERGY AGENCY, Conducting Computer Security Assessments at Nuclear Facilities, Non-serial Publications, IAEA, Vienna (2016). (a revision of this publication is in preparation.)

[14] INTERNATIONAL ATOMIC ENERGY AGENCY, Code of Conduct on the Safety and Security of Radioactive Sources, IAEA, Vienna (2004).

[15] INTERNATIONAL ATOMIC ENERGY AGENCY, Security of Radioactive Material in Use and Storage and of Associated Facilities, IAEA Nuclear Security Series No. 11-G (Rev. 1), IAEA, Vienna (2019).

[16] EUROPEAN COMMISSION, FOOD AND AGRICULTURE ORGANIZATION OF THE UNITED NATIONS, INTERNATIONAL ATOMIC ENERGY AGENCY, INTERNATIONAL LABOUR ORGANIZATION, ORGANISATION FOR ECONOMIC CO-OPERATION AND DEVELOPMENT NUCLEAR ENERGY AGENCY, PAN AMERICAN HEALTH ORGANIZATION, UNITED NATIONS ENVIRONMENT PROGRAMME, WORLD HEALTH ORGANIZATION, Radiation Protection and Safety of Radiation Sources: International Basic Safety Standards, IAEA Safety Standards Series No. GSR Part 3, IAEA, Vienna (2014), https://doi.org/10.61092/iaea.u2pu-60vm

[17] INTERNATIONAL ATOMIC ENERGY AGENCY, Categorization of Radioactive Sources, IAEA Safety Standards Series No. RS-G-1.9, IAEA, Vienna (2005).

[18] INTERNATIONAL ATOMIC ENERGY AGENCY, Security Management of Radioactive Material in Use and Storage and of Associated Facilities, IAEA Nuclear Security Series No. 43-T, Technical Guidance, IAEA (2022).

[19] US DEPARTMENT OF ENERGY, NATIONAL NUCLEAR SECURITY ADMINISTRATION, OFFICE OF RADIOLOGICAL SECURITY, Cybersecurity Best Practices for Users of Radioactive Sources, ORS, Washington (2022).

[20] INTERNATIONAL ORGANIZATION FOR STANDARDIZATION AND INTERNATIONAL ELECTROTECHNICAL COMMISSION, Information technology — Security techniques — Information security management systems — Overview and vocabulary, ISO/IEC 27000:2018, ISO, Geneva (2018).

[21] INTERNATIONAL ORGANIZATION FOR STANDARDIZATION AND INTERNATIONAL ELECTROTECHNICAL COMMISSION, Information security, cybersecurity and privacy protection — Guidance on managing information security risks, ISO/IEC 27005:2022, ISO, Geneva (2022).

[22] CAPABILITY MATURITY MODEL INTEGRATION INSTITUTE, Capability Maturity Model Integration V2.0, CMMI Institute, Pittsburgh (2018), https://cmmiinstitute.com/cmmi

[23] INTERNATIONAL ORGANIZATION FOR STANDARDIZATION AND INTERNATIONAL ELECTROTECHNICAL COMMISSION, Information security, cybersecurity and privacy protection — Information security controls, ISO/IEC 27002:2022, ISO, Geneva (2022).

[24] US DEPARTMENT OF ENERGY, NATIONAL NUCLEAR SECURITY ADMINISTRATION, OFFICE OF RADIOLOGICAL SECURITY, Cybersecurity Procurement Requirements for ORS-Provided Security Systems, ORS, Washington (2022).

[25] INTERNATIONAL ELECTROTECHNICAL COMMISSION, Security for industrial automation and control systems – Part 4-2: Technical security requirements for IACS components, IEC 62443-4-2:2019, IEC, Geneva (2019).

[26] INTERNATIONAL ORGANIZATION FOR STANDARDIZATION AND INTERNATIONAL ELECTROTECHNICAL COMMISSION, Information security, cybersecurity and privacy protection – Evaluation criteria for IT security – Part 1: Introduction and general model, ISO/IEC 15408-1:2022, ISO, Geneva (2022).

[27] UL, Software Cybersecurity for Network-Connectable Products, Part 1: General Requirements, Standard 2900-1, Edition 1, UL Standard, Northbrook (2017).

[28] INTERNATIONAL ATOMIC ENERGY AGENCY, Nuclear Security Recommendations on Physical Protection of Nuclear Material and Nuclear Facilities (INFCIRC/225/Revision 5), IAEA Nuclear Security Series No. 13, IAEA, Vienna (2011), https://doi.org/10.61092/iaea.ko2c-dc4q

[29] NATIONAL INSTITUTE OF STANDARDS AND TECHNOLOGY, Computer Security Resource Center, Security and Privacy Controls for Information Systems and Organizations, NIST SP 800-53 Revision 5, NIST, Gaithersburg, (2020), https://doi.org/10.6028/NIST.SP.800-53r5

[30] INTERNATIONAL ATOMIC ENERGY AGENCY, Computer Security Approaches to Reduce Cyber Risks in the Nuclear Supply Chain, Non-serial Publications, IAEA, Vienna (2022).

[31] INTERNATIONAL ATOMIC ENERGY AGENCY, Handbook on the Design of Physical Protection Systems for Nuclear Material and Nuclear Facilities, IAEA Nuclear Security Series No. 40-T, IAEA (2021).

BIBLIOGRAPHY

Computer security resources for facilities using radioactive material

Center for Internet Security, www.cisecurity.org/

Common Weakness Enumeration, CWE Top 25 Most Dangerous Software Weaknesses, http://cwe.mitre.org/top25/

Cyber Security & Infrastructure Security Agency, www.cisa.gov/resources-tools/resources

Department of Homeland Security, Cybersecurity, www.dhs.gov/topic/cybersecurity

Government of Canada, Cyber Security, www.canada.ca/en/services/defence/cybersecurity.html

Institute of Electrical and Electronics Engineers, IEEE Internet Technology Policy Community White Paper, Internet Of Things (IOT) Security Best Practices, https://internetinitiative.ieee.org/images/files/resources/white_papers/internet_of_things_feb2017.pdf, http://internetinitiative.ieee.org/

MITRE, Cybersecurity, www.mitre.org/focus-areas/cybersecurity

NIST Cybersecurity Framework, www.nist.gov/cyberframework/

SANS Institute, CIS Controls v8, www.sans.org/critical-security-controls

United States Nuclear Regulatory Commission, Backgrounder on Cyber Security, www.nrc.gov/reading-rm/doc-collections/fact-sheets/cyber-security-bg.html - require

ANNEX I: KNOWN CYBER-ATTACKS AGAINST FACILITIES USING, HANDLING AND/OR STORING RADIOACTIVE MATERIAL

This annex provides a summary of known cyber-attacks in digital systems performing functions important to the security of radioactive material. The examples are not intended to represent an exhaustive list of cyber-attacks, nor are they intended as a recipe for gaining access to radioactive material or for sabotaging associated activities. Rather, they are a starting point for licensees and States, to assist them in the development of plans in order to address threats acting in a dynamic, rapidly changing cyber environment. Personnel with computer security responsibilities will need to be able to imagine the different ways in which adversaries might act and how computer security measures could counter their actions.

Table I–1 provides examples of some publicly available information on cyber-attacks that have occurred in relation to the computer security of the PPS, medical devices or networks that could be present at facilities using, handling and/or storing radioactive material. These examples demonstrate the types of threat against which computer security programmes need to prepare their operational technologies (e.g. physical protection systems).

TABLE I–1. EXAMPLES OF CYBER-ATTACKS AGAINST FACILITIES USING, HANDLING AND/OR STORING RADIOACTIVE MATERIAL

Place and year of the incident	**Example**	**Threat**
Canada 2020	A ransomware attack on a computer system that stores confidential medical data for Saskatchewan residents affected approximately 40 patients. These patients were getting cancer treatment in Saskatoon and Regina (see Ref. [I–1]).	Ransomware
Norway 2019	A ransomware attack, known by investigators as LockerGoga, entered the Norsk-Hydro company's computer system in Oslo, Norway. The ransomware encrypted the computer system's files and halted operations in multiple locations globally, forcing Norsk Hydro to suspend normal operations and switch to manual operations, where necessary (see Ref. [I–2]).	Ransomware
United States 2019	A ransomware infection at an undisclosed US Coast Guard facility led to a disruption of cameras and physical access control systems, and to a loss of the key process control monitoring systems. The attacker gained access to significant IT network files of the enterprise, disrupting the entire network beyond the facility (see Ref. [I–3]).	Ransomware
United States 2019	The US Food and Drug Administration (FDA) issued a safety communication for healthcare organizations, IT professionals, device manufacturers and patients to warn of a cybersecurity vulnerability known as URGENT/11. According to FDA officials, the risk was that URGENT/11 could be exploited by a remote attacker and could subsequently pose safety and security risks for connected medical devices and hospital networks (see Ref. [I–4]).	Protocol

TABLE I–1. EXAMPLES OF EVOLVING CYBER-ATTACKS AGAINST AT FACILITIES USING, HANDLING AND/OR STORING RADIOACTIVE MATERIAL (cont.)

Place and year of the incident	Example	Threat
Global 2018	First identified in January 2015, Orangeworm has conducted targeted attacks against organizations in industry as part of a larger supply chain attack to reach its victims, including healthcare providers, pharmaceuticals, IT solution providers for healthcare and equipment manufacturers that serve the healthcare industry. The likely purpose of Orangeworm is corporate espionage (see Ref. [I–5]).	Supply chain risk management
Global 2017	The worldwide cyber-attack by WannaCry ransomware, which targeted the Microsoft Windows operating system and encrypting data, demanding ransom payments, was unprecedented in scale. It infected more than 230 000 computers in over 150 countries (see Ref. [I–6]).	Ransomware
Global 2016	The Internet of Things (IoT) has been used to create large scale botnets (i.e. networks of devices infected with self-propagating malware) that can execute crippling distributed denial-of-service attacks. IoT devices affected were mainly home routers, network enabled cameras and digital video recorders (see Ref. [I–7]).	Distributed denial-of-service
United States 2015	TrapX Security discovered that attackers had compromised a C-ARM X ray system through email and then pivoted off the X ray device to each hospital subnetwork that it was connected to as it moved around to patient locations throughout the facility, which acted as a host for this advanced persistent threat. The threat was immune to regular IT security deployments (see Ref. [I–8]).	Protocol

REFERENCES TO ANNEX I

[I–1] ZAKRESKI, D., Ransomware attack on eHealth forces 31 cancer patients to re-schedule radiation treatment; six patients booked for chemotherapy also affected, CBC News (2020), www.cbc.ca/news/canada/saskatoon/ransomware-attack-ehealth-cancer-patients-1.5428346

[I–2] ZORZ, Z., Norsk Hydro cyber attack: What happened? Help Net Security (2019), www.helpnetsecurity.com/2019/03/20/norsk-hydro-cyber-attack/

[I–3] CIMPANU, C., US Coast Guard discloses Ryuk ransomware infection at maritime facility, ZDNET (2019), www.zdnet.com/article/us-coast-guard-discloses-ryuk-ransomware-infection-at-maritime-facility/

[I–4] MILIARD, M., URGENT/11: FDA issues alert for cyber vulnerability that threatens medical devices, networks, Healthcare IT News (2019), www.healthcareitnews.com/news/urgent11-fda-issues-alert-cyber-vulnerability-threatens-medical-devices-networks

[I–5] Threat Hunter Team: Symantec, New Orangeworm attack group targets the healthcare sector in the U.S., Europe, and Asia, Symantec (2018), https://symantec-enterprise-blogs.security.com/blogs/threat-intelligence/orangeworm-targets-healthcare-us-europe-asia

[I–6] WIKIPEDIA, WannaCry ransomware attack (2023), https://en.wikipedia.org/wiki/WannaCry_ransomware_attack

[I–7] CYBERSECURITY & INFRASTRUCTURE SECURITY AGENCY, Heightened DDoS Threat Posed by Mirai and Other Botnets (2017), www.cisa.gov/news-events/alerts/2016/10/14/heightened-ddos-threat-posed-mirai-and-other-botnets

[I–8] TrapX RESEARCH LABS, MEDJACK.4 Medical Device Hijacking, TrapX Investigative Report, TrapX Security, Inc, San Jose (2018).

ANNEX II: US OFFICE OF RADIOLOGICAL SECURITY BEST PRACTICES (AS OF 2022) FOR CYBERSECURITY FOR USERS OF RADIOACTIVE SOURCES

Designing and implementing effective security measures for radioactive sources entails consideration of security, response, and administrative and management elements to ensure that all work together to have a successful computer security programme that prevents the theft of radioactive sources or the sabotage of Internet protocol (IP) based equipment. Table II–1 highlights selected computer security controls (i.e. measures) from the US Department of Energy, National Nuclear Security Administration, Office of Radiological Security [II–1]. Such controls could be considered as part of a computer security programme. Individual sites might already have some of these controls in place, or they could be implemented with assistance from the IT department at the site, or with the help of outside computer security expertise. The designation of the controls as either 'initial', 'intermediate' or 'advanced' are to be understood as simple guidelines. In the case of some sites, initial measures could be considered adequate. In other cases, however, the implementation of further intermediate and advanced controls could be considered necessary as the site's computer security programme matures. The best practices listed in Table II–1 are presented in the same order as those listed in Ref. II–1.

TABLE II–1. COMPUTER SECURITY BEST PRACTICES [II–1]

Control	Initial	Intermediate	Advanced
Designated computer security personnel	The site management designates a person or member of the personnel as being responsible for computer security.	Computer security responsibilities are enumerated, with specific descriptions provided.	Computer security responsibility and accountability are defined and allocated.
User accounts	Strict user accounts, with limited role based permissions, are made mandatory. The principle of 'least privilege' is used for access to systems.	Each user has his or her individual account with the appropriate role based permissions.	Role based accounts with appropriate minimal permissions that are shared among similar users (e.g. medical staff, response forces, alarm monitoring personnel) are allocated and in use.
Password management	Strong, complex password management is enforced.	Passwords are to be changed periodically. Passwords for shared accounts are changed when personnel change roles.	Strong passwords are enforced. These passwords cannot be the default passwords for devices.
Two-factor authentication	In addition to strong, complex password management, additional login authentication (e.g. RSA tokens) is employed. Multifactor authentication is used.	Two-factor authentication is necessary for all user accounts on all systems, where practical.	Two-factor authentication is necessary for administrator accounts on all systems, where practical. Two-factor authentication is implemented on higher consequence systems only.

TABLE II–1. COMPUTER SECURITY BEST PRACTICES [II–1] (cont.)

Control	Initial	Intermediate	Advanced
Unnecessary accounts and software	Unnecessary accounts, software and processes are removed.	Each account is justified or otherwise periodically removed. Each individual software and process is justified or removed.	Unused user accounts are disabled or removed. During installation of software, unnecessary software and processes are removed.
Anti-malware	Anti-malware software is installed, and the personnel ensure that it is kept current.	Anti-malware software is updated upon alert of signature availability for current threats. The system is scanned promptly after updates, and reports are regularly reviewed. Higher frequency of scheduled scanning is expected. Stronger anti-malware defences (e.g. application whitelisting) are deployed and operational.	Anti-malware software is installed and updated periodically. Information on whether anti-malware signatures are current is used as a condition to connect to organization networks, where practical. Systems are scanned periodically.
Acceptable computer use policy	Management ensures that the site has an acceptable use policy for personnel using the facility's computer security resources.	Users regularly receive reminders and refresher training.	The site has an acceptable use policy that is signed by users before allowing access to systems.
Baseline inventory of hardware and software	Management establishes a baseline to identify all equipment, update documentation to correspond to the physical implementation of the system and implement configuration and change management processes.	A comprehensive configuration and change management system is established that includes the installation of hardware, serial numbers, software, and firmware (with version numbers), which are updated as part of any change.	Configuration and change management information is updated after major changes (e.g. hardware or software updates and additions). An approval process exists for configuration changes. Configuration and change management information is available.
Patch and firmware integrity	Patches and firmware are to be derived from authorized vendors.	The integrity of patches and firmware is verified (e.g. through a checksum comparison). In cases that this is not possible, patches and firmware updates are applied in a sandbox environment, where practical.	Patches and firmware are obtained from original product vendors.
Firmware updates	Network switches, alarm panels, access control devices, the computer basic input output system (BIOS), digital cameras and other components need to be patched for the current firmware version.	Information sources (e.g. vendors, computer emergency response teams (CERTs)) are monitored for security significant firmware updates. When released, such updates are applied as soon as possible or compensatory measures are identified and applied.	Equipment firmware updates are applied periodically in accordance with local procedures. Some medical devices or other identified devices could be in need of additional testing and accreditation to ensure that updates do not have a negative impact on device operation.
Enterprise class hardware	Enterprise class hardware is purchased and used instead of consumer class components that are meant for home or small office use.	All equipment is enterprise class hardware, with support provided by authorized vendors.	

TABLE II–1. COMPUTER SECURITY BEST PRACTICES [II–1] (cont.)

Control	Initial	Intermediate	Advanced
Authorization to introduce new software and firmware	All software and firmware upgrades are to be performed by authorized system administrators or managers only.	Technical controls (e.g. application whitelisting) are used to enforce administrative controls.	Administrative controls are used to permit only authorized users to introduce new software and firmware (e.g. updates, installation of new applications).
Web browser configuration	Alarm management software is often in need of web browsers and dedicated email accounts. The browser is to be configured to limit access to non-system related sites.	Technical measures are used to limit active content executed by the web browser. Web browsers are only used to access websites relevant to system function. If not required, the web browser is disabled or removed, if practical.	Web browsers on sensitive digital assets are configured to limit access to non-system related websites or are limited by external means (e.g. network connectivity, firewall)
Physical hardening	Physical hardening of the intrusion detection system and the access control system of host computer locations, workstations, wiring closets and on-site central monitoring stations is undertaken.	Sensitive digital assets are considered to be within the physical protection system, and thus their protection is commensurate with their overall consequence of compromise.	
Port scanning	Port scanning is performed for all PPS components that connect to the network and communication infrastructure.	Port scanning of digital assets is undertaken, where practical, and non-destructive measures are used to verify open services against a baseline on a periodic basis.	
Patching vulnerabilities	Vulnerabilities on all ports and associated services are patched.	Information sources (e.g. vendor, CERT) are monitored for security significant system and application patches. When released, such patches are applied as soon as possible or compensatory measures are identified and applied.	Systems and applications are patched periodically.
Unnecessary ports	Disable all unnecessary ports and associated services through hardware and software hardening	Disabled network services and physical ports are monitored and reviewed on a periodic basis to ensure they remain disabled.	Unnecessary network services and physical ports are disabled on system installation.
Mobile device management (MDM)	Use MDM for the administration of mobile devices accessing facility networks.	MDM is an active and mandatory programme.	
Recovery after a computer security incident	Management ensures that the site has a strategy for the development and implementation of plans, processes and procedures for the recovery and full restoration of any capabilities or services that are impaired as a result of a cyber-attack, all in a timely manner.	The site management has a documented and exercised recovery plan for cyber-attacks.	
Personnel awareness of phishing	Management ensures that the site has an active security awareness programme, potentially including phishing campaigns to test the security awareness of personnel.	Active testing occurs to measure users' abilities to detect phishing attempts.	Users are provided with phishing awareness material and training.

TABLE II–1. COMPUTER SECURITY BEST PRACTICES [II–1] (cont.)

Control	Initial	Intermediate	Advanced
Firewall attack detection, logging and alerting features	Management enables built-in firewall attack detection, and the logging and alerting features that already exist in most modern firewalls.	Firewall logs are reviewed periodically, and processes exist to respond to alerts that are generated.	The firewall is configured for attack detection, with logging and alerting features.
Enforce network traffic flows	Network traffic flows in existing firewalls are enforced.	The firewall configuration is reviewed and updated periodically on the basis of current required traffic flows (e.g. rules that are no longer necessary are removed).	The firewall is used to allow only required communication and to block all other communication attempts.
De-militarized zone (DMZ)	Utilize the existing firewall DMZ, as applicable (e.g. drop boxes, servers for the domain name system, web servers).	A DMZ is implemented with two firewalls using diverse technology platforms.	
Port security	Port security is enabled on network switches, unused interface ports are disabled and administrative access is restricted.	Port security is enabled and actively controlled, administrative access to change port security is enforced.	Port security is enabled at installation of equipment. Physical security measures are used to secure switches.
Access control lists (ACLs) and administrative access	ACLs are created and administrative access is restricted.	ACLs are created to enforce role based permissions for all users.	Administrative access is restricted to authorized personnel.
Minimize network connection	The system is air gapped, if possible, or at least the number of perimeter interconnections is minimized to provide network isolation, where feasible.	The system is air gapped (with data diodes included) from all other networks on the basis of a defined trust model.	Network segmentation is implemented within a zoned configuration. Wireless access is not permitted to networks containing sensitive digital assets. All network interconnections are identified and security measures reused to provide network decoupling.
Zone architecture	Multizone network security architecture is configured to isolate security protection components into logical groups. New, transparent mode firewalls are added, where needed, while the minimum required traffic flows between the zones are enforced.	Formal and documented defensive computer security architecture is implemented and reinforced through detection, delay and response measures.	Ad hoc zone architecture — A documented boundary control between the corporate network and sensitive digital assets is identified.
Thin clients	Thin client network terminals are enforced instead of Windows workstations, where possible, to reduce the attack surface, patch requirements, and ensure total cost of ownership.	Thin clients are in limited use.	In large or multiple target facilities, thin clients are used strategically as part of the defensive security architecture to reduce the overall attack surface.
Traffic encryption	Traffic encryption for communication over any external networks or telecommunications circuits is enforced.	Traffic is strategically encrypted in internal networks to ensure the protection of sensitive information.	Traffic is encrypted over external networks and telecommunications circuits.

TABLE II–1. COMPUTER SECURITY BEST PRACTICES (cont.)

Control	Initial	Intermediate	Advanced
IP address management	Redundant, non-routable, static IP dedicated networks are employed in the core design.	Assigned IP addresses are considered in the specifications of security policies to support protection and detection.	IP addresses are assigned on a per asset basis to support monitoring.
Intrusion detection	Intrusion detection capability is added to analyse network traffic and identify alerts that are generated for attempted cyber-attacks or suspicious packets and payloads.	The intrusion detection system can take automated prevention actions against known cyber-attacks without the need for human assessment.	Intrusion detection is performed by automated systems, and processes exist to review and respond to any alerts that are generated. Ad hoc intrusion detection is a defined process that is employed when there is another indicator of compromise.
Testing for cyber vulnerabilities	Prior to deployment, new equipment and components are thoroughly tested for cyber vulnerabilities.	A vulnerability assessment (e.g. penetration test) is performed by a competent professional.	Prior to deployment, new equipment and components are tested for cyber vulnerabilities through an automated tool.

REFERENCE TO ANNEX II

[II–1] US Department of Energy, National Nuclear Security Administration, Office of Radiological Security (ORS), Cybersecurity Best Practices for Users of Radioactive Sources, ORS, Washington, DC (2018).

ANNEX III: SECURITY EQUIPMENT, REMOTE MONITORING AND ADDITIONAL CONSIDERATIONS FOR INFORMATION AND COMPUTER SECURITY

An effective physical security programme entails the integration of people, procedures and equipment in order to protect radioactive sources from theft, or from criminal or other intentional unauthorized attacks. This integration includes computer security controls to ensure that security protection elements are not compromised by a cyber-attack. The blending of physical protection systems (PPSs) with information technology is advancing at such a rapid pace that the two can no longer be viewed independently. In addition, security systems are evolving from stand-alone hardwired devices to network based devices where both power and data are provided by a single Ethernet cable. Annex III examines physical security devices and their communication with alarm monitoring stations, along with the computer security concerns associated with these devices.

System components are increasingly being reviewed by computer security professionals for vulnerabilities, including hardwired and proprietary code, IP based communication and non-proprietary protocols. Facilities using, handling, and/or storing radioactive sources need to review their security systems to see what parts of systems have networks for communication and services (e.g. security cameras in public spaces with USB ports or other interfaces), as these are potential adversary pathways for cyber-attacks. Hardware and software hardening — for example, removing unnecessary programs and services and blocking unnecessary access ports, such as USB drives and firewall ports — can reduce these potential adversary pathways.

All security systems contain some form of intrusion detection system, access control systems and a method for monitoring alarm states either on the site, off the site or, in many cases, both on and off the site. The main security system, and network related components and capabilities that are potentially vulnerable to cyber-attacks, include:

- Alarm concentrators and panels, which communicate to the host using various communications protocols over Ethernet, cellular or other combination of communication means;
- Global system communication channels for alarm transmissions through mobile communication, a general packet radio service or through a radio frequency;
- Analog cameras, which are being replaced by IP cameras in cases where power over Ethernet (PoE) is becoming even more commonplace;
- Network infrastructure, mainly comprising switches, midspans (PoE injectors or switches), repeaters, routers (wired and wireless) and firewalls, and in some installations, wireless access points (WAP), as well as IP addressable access control and alarm keypads (biometrics, proximity cards, pin keypads);
- Application servers cloud based services, such as infrastructure as a service (IaaS), software as a service (SaaS) and platform as a service (PaaS);
- Remote access.

Most alarm sensor devices (e.g. motion sensors, balanced magnetic switches) cannot currently be attributed an IP address and are hardwired into alarm panels; but the security industry is quickly moving towards devices that can be attributed IP addresses, in some cases offering wireless connectivity. A move to sensor devices with IP addresses will broaden the cyber-attack surface by including these security system components. In addition, current systems generally use existing telephones or radios for communication, although these telephones and radios are not integrated into the security system. As current and future systems are built to take advantage of the cost savings provided by IP networks, such as voice over Internet protocol (VoIP), the use of these converged mediums introduce other attack surfaces for adversaries to target.

Figure III–1 depicts an example of a typical blood irradiator room (i.e. containing a high activity radioactive source) with physical security devices installed in accordance with IAEA Nuclear Security Series No. 11-G (Rev. 1), Security of Radioactive Material in Use and Storage and of Associated Facilities [III–1]. The diagram also depicts a remote monitoring system device, which is a stand-alone security system that is independent of the rest of the physical security devices that provides protection against an insider threat.

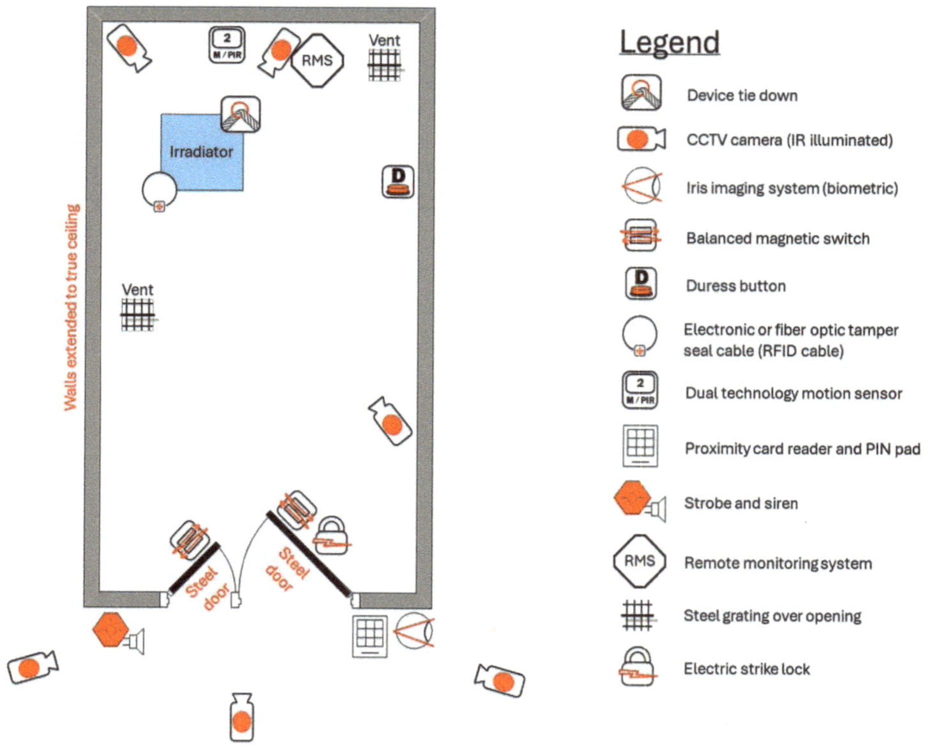

FIG. III–1. Example installation layout of physical security devices in a blood irradiator room.

Physical security devices increasingly use network based communication that can increase a site's vulnerability to a cyber-attack. IT and computer security personnel can help identify which of the physical security devices use site networks or other IP based communication. These networks or IP based communications could potentially be an additional attack surface for an adversary. Isolation and network segmentation of security systems on-site, through measures such as virtual local area networks (VLANs), can be a very important measure to address many potential security issues.

Access control measures enable facilities to implement systems and procedures to deny the access of unauthorized persons to target rooms while allowing the access of authorized individuals. A biometric access control device on the access door(s) to the device room could also be installed to prevent authorized persons from providing their access credentials (e.g. badge, PIN code) to an unauthorized person. Common biometric systems include fingerprint or handprint readers, hand geometry readers, iris readers, retinal readers and hand vascular pattern recognition readers. Other access control options, such as PINs, proximity cards, or a PIN and a proximity card might be more appropriate instead of a biometric device, depending on the site's security needs.

Figure III–2 depicts an example of a typical intrusion detection system designed to detect an adversary before they reach the radioactive source. Security vendors can recommend the best type of alarm sensors for inclusion in an intrusion detection system, but sensors are to be of commercial quality similar to what would be used in a bank or other facility with high value assets. Typical home security alarm sensors are to be avoided. Tamper indicating devices, such as radio frequency identification tags could be included as components of an intrusion detection system. Computer security concerns include the method of signal communication between the intrusion detection system alarm panel and the monitoring stations.

The main purpose of the video surveillance system is to assess the alarm and have sufficient camera resolution and lighting to provide a clear picture to the monitoring station, which shows that there is unauthorized activity occurring in relation to the source device or sources. Video surveillance systems are increasingly network based and therefore vulnerable to cyber-attack.

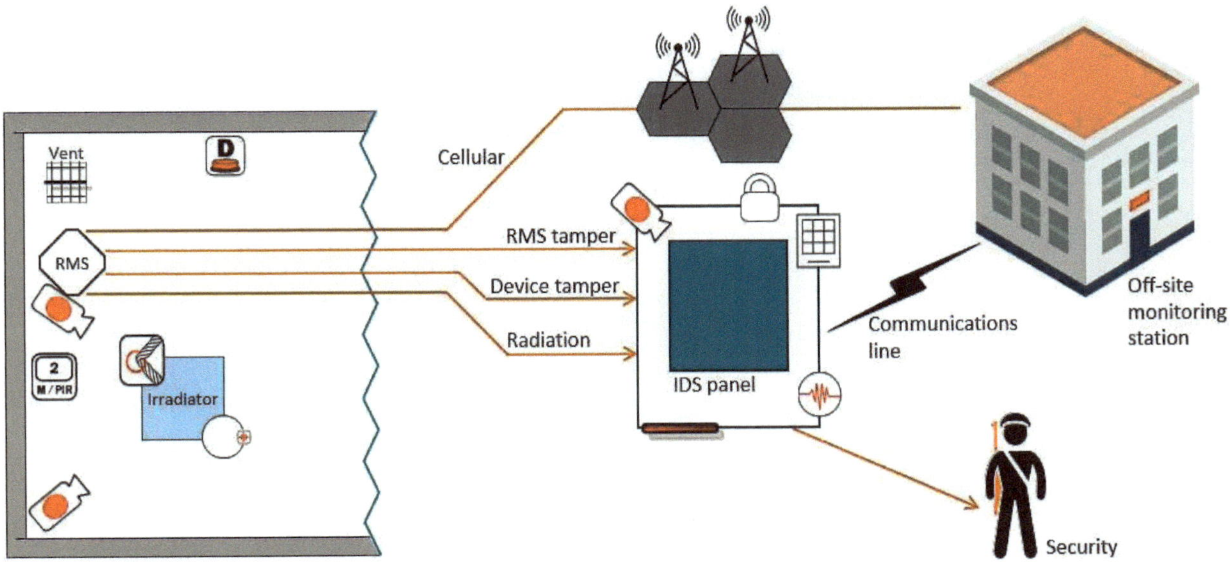

FIG. III–2. Intrusion detection system.

Figure III–3 depicts an example of a typical remote monitoring system (RMS) with a separate intrusion detection system that incorporates an active tamper indicating device on the asset and a radiation detector that is used for security purposes. The system is to have redundant video surveillance focused on the device, providing protection against insiders since these alarms and videos are always on, except during authorized maintenance and testing (i.e. the device user cannot disable these alarms). The RMS provides the capability for remote (off-site) monitoring of crucial alarms and videos to rapidly provide information to armed responders who will assist them in responding to a theft. In most cases, the means of communication will be over the Internet. In terms of data communication to an off-site monitoring station, the selection of the technical means of transmitting RMS data from a building or site to an alarm monitoring station is dependent on several factors, including the state of the communications infrastructure, the reliability of the telephone systems, and the capability and availability of cellular data networks (e.g. the global system for mobile communications, general packet radio service, 3G, 4G, radio frequency capability). The use of telephone lines with autodialers is not an option that is favoured since this method does not provide frequent enough polling to ensure the continuous monitoring of alarm data transmission; nor does it ensure that the loss of communication is detected in a timely manner. In some instances, two means of alarm transmission, such as Internet and cellular, could be necessary to ensure reliable alarm transmission with polling capability and timely detection of the loss of communication.

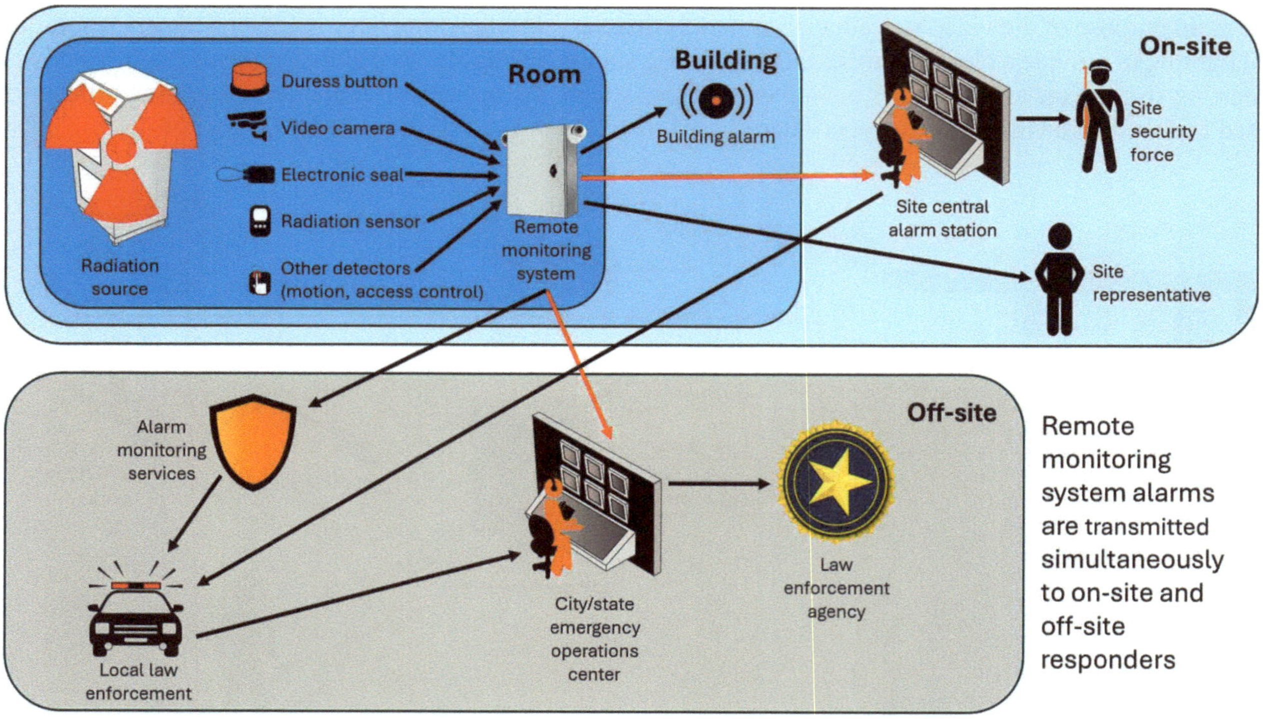

FIG. III–3. Remote monitoring system.

Site management could choose to provide or significantly enhance an on-site alarm monitoring station on the basis of the role that the station plays in the initiation of a response and its role in the implementation of the site protection strategy. In most cases, a site can build upon existing on-site monitoring capabilities. Since a site is also to have off-site alarm monitoring, small sites with a limited on-site security force (e.g. unarmed) might consider the expenditure of resources for the construction of a new monitoring station to be unwarranted. Sites with highly attractive radioactive material, a capable on-site armed response force, short adversary timelines, or where the on-site alarm monitoring station is crucial to implementing the site protection strategy, could consider establishing a hardened or protected on-site monitoring station. The method of alarm communication will often be over the site's network.

REFERENCE TO ANNEX III

[III–1] INTERNATIONAL ATOMIC ENERGY AGENCY, Security of Radioactive Material in Use and Storage and of Associated Facilities, IAEA Nuclear Security Series No. 11-G (Rev. 1), IAEA, Vienna (2019).

IAEA
International Atomic Energy Agency

CONTACT IAEA PUBLISHING

Feedback on IAEA publications may be given via the on-line form available at:
www.iaea.org/publications/feedback

This form may also be used to report safety issues or environmental queries concerning IAEA publications.

Alternatively, contact IAEA Publishing:

Publishing Section
International Atomic Energy Agency
Vienna International Centre, PO Box 100, 1400 Vienna, Austria
Telephone: +43 1 2600 22529 or 22530
Email: sales.publications@iaea.org
www.iaea.org/publications

Priced and unpriced IAEA publications may be ordered directly from the IAEA.

ORDERING LOCALLY

Priced IAEA publications may be purchased from regional distributors and from major local booksellers.

25-01271E